SpringerBriefs in Applied Sciences and Technology

For further volumes:
http://www.springer.com/series/8884

Alireza Bahadori · Malcolm Clark
Bill Boyd

Essentials of Water Systems Design in the Oil, Gas, and Chemical Processing Industries

Springer

Alireza Bahadori
Malcolm Clark
Bill Boyd
School of Environment,
    Science and Engineering
Southern Cross University
Lismore, NSW
Australia

ISSN 2191-530X   ISSN 2191-5318 (electronic)
ISBN 978-1-4614-6515-7   ISBN 978-1-4614-6516-4 (eBook)
DOI 10.1007/978-1-4614-6516-4
Springer New York Heidelberg Dordrecht London

Library of Congress Control Number: 2013932784

Printed on acid-free paper

Springer is part of Springer Science+Business Media (www.springer.com)

Preface

Water supply systems are crucial in supporting industrial oil, gas, and other chemical processing systems. Reliability of supply and cost of water to such industries is important for both the sustainable management of such industries and for the provision of a supply market. This book, therefore, overviews and introduces the technical matters related to the process design and selection of water supply systems used in such industries. In doing so, it provides an introduction to the field of industrial processing water supply management, and a frame for further engagement in the detailed literature of this field.

Both process design and process selection of the many water supply systems used in oil, gas, and chemical processing industries are crucial for the maintenance of existing facilities and the design of next-generation processing industries. Simply put, oil, gas, and most chemical processing plants cannot function without water-based utility systems. Although the importance of these systems is not usually contested, expansion or upgrade expenditures of these operations are often avoided, because no direct payback can be assigned to any utility capital expenditures.

The cost of supplying water for steam, cooling, and processing varies greatly, depending on the water source. Water typically comes from sources such as on-site groundwater wells, surface water, or off-site providers. These supplies often have flow limit restrictions, and the purchasing water can be expensive. There may also be additional regulations enforced when demand exceeds permitted limits. Moreover, the cost for raw water treatment (chemical additives, softeners, and flocculants), sludge disposal, pumping, and other processing, rises with increased water demand.

Water use at a plant can increase for many reasons; hence plant expansions and unit conversions can impact utility systems by boosting flows and contaminant loading. In addition, new and modified units may contribute to increased storm water runoff, and more stringent quality specifications may also increase water demand from increased washing/treatment steps.

The aim of this book is to provide an overview of the main technical points related to the process design and selection of water supply systems used in the oil, gas, and chemical processing industries. This overview is framed around four systems that, together, provide an integrated industrial water management system.

- Water treatment system
- Raw water and plant water system
- Water pollution
- Fire water distribution and storage facilities

There is a direct relationship between water demand and flows to water treatment. Consequently, many water treatment and wastewater units are designed for peak flows only experienced during storm conditions. Treatment costs during these peak flow conditions can climb exponentially from increased pumping, aeration demands, sludge management, and solids disposal requirements. Most importantly, additional water use reduces treatment capacity during peak flows often resulting in the need for additional storage capacity to dampen these peaks.

The field of process water design is broad, and contains a wide range of subjects, each of paramount importance including raw water treatment and recovery systems. The treatment of both water and wastewater involves a sequence of treatment steps. All water and wastewater treatment processes involve the separation of solids from water in at least some part of the operation and removal of biochemical oxygen demand (BOD) to some extent. The end of pipe treatment sequence can be divided into the following elements: primary or pretreatment; intermediate treatment; secondary treatment; and tertiary treatment plus ancillary, sludge dewatering, and disposal operations.

Optimizing the performance of individual unit operations, such as gravity separator, dissolved air flotation, biological treatment, etc., can best be achieved if:

- the properties of influent streams are considered;
- the chemical principles that are used in solids pretreatment are understood;
- the variety of chemicals available for solids treatment is recognized;
- the properties of effluent water are established based on the local environmental regulations and final disposal; and
- the protocols for quantifying results are identified.

Effluent wastewaters are a combination of the liquid and water-carried wastes from buildings, industrial plants, plus groundwater, surface water, or storm water. Wastewater may be grouped into the following classes [5–8]:

Class 1   Effluents that are non-toxic, and not directly polluting but liable to disturb the physical nature of the receiving water, may be improved by physical means.

Class 2   Effluents that are non-toxic, but polluting because they have an organic content with high oxygen demand, may be treated for removal of objectionable characteristics by biological methods.

Class 3   Effluents that contain toxic materials, and therefore are polluting, may be treated by chemical methods.

Class 4   Effluents that are polluting because of organic content with high oxygen demand and, in addition are toxic, may require a combination of chemical, physical, and biological processes.

The final release of effluents and surface water drainage to the broader environment is subject to the approval of environmental scientists and experts, a factor that must be borne in mind in the early stages of design. In general, the aim of any drainage/effluent disposal system should be to segregate uncontaminated water from contaminated water or effluents and to segregate different types of effluents in order to reduce the size, complexity, and costs of any treatment units which may be required for handling the contaminated water and effluents before they are discharged from oil, gas, and chemical processing plants.

Dr. Alireza Bahadori
Dr. Malcolm Clark
Prof. Bill Boyd

Contents

Chapter 1
Water Treatment Systems

Keywords Source water · Water treatment · Treatment process selection · Water quality · Boiler water · Total suspended solids · Total dissolved solids · Silica · Deposits in water systems

1.1 Introduction

The increasing energy demand over the last decades has resulted in a corresponding growth and expansion in the processing of crude petroleum, natural gas and other hydrocarbon resources. Water treatment or the purification of water varies as to the source and kinds of water. Type of water treatment depends on the quality of the source water and the quality desired in the finished water. Adequate information on the source water is thus a prerequisite for design. This includes analysis of the water and, where the supply is non-uniform, the ranges of the various characteristics.

The proliferation of oil, gas, petrochemical and chemical plants, combined with increasingly stringent discharge limit requirements for effluents from these facilities, underscores the need for improving existing pollution control technologies or developing new and improved approaches for minimizing the pollution potential in the oil, gas, petrochemical and chemical sector [1, 2].

For example, in petroleum industry, nearly all crude oils contain some basic sediment and water (BS&W), which is generally composed of a mixture of water, iron rust, iron sulfides, clay, sand and particulate contaminates produced with the crude oil or picked up in transit. Part of the BS&W is charged to the crude oil unit and may settle out in the desalter, entering the oily water sewer system along with the desalter effluent. Similarly, water in contact with process streams, originating from steam stripping, crude oil washing, some chemical oil treatment processes etc., may contain variable amounts of oil [3, 4].

Suspended solids, biodegradable organics, nutrients, refractory organics, heavy metals, dissolved inorganic solids, pathogens, soluble material such as ammonium sulfide, phenols, thiophenols, organic acids and inorganic salts such as sodium

A. Bahadori et al., *Essentials of Water Systems Design in the Oil, Gas, and Chemical Processing Industries*, SpringerBriefs in Applied Sciences and Technology, DOI: 10.1007/978-1-4614-6516-4_1, © The Author(s) 2013

chloride are the important contaminants which may be found in the oil, gas and chemical processing industry's utility waters and wastewaters [5, 6].

Relevant to the above-mentioned facts, these five categories of water may need different treatments, and for this reason, water streams are often kept segregated in a modern refinery to reduce the cost of water treatment facilities.

Water treatment requirements for oil, gas, chemical processing industries and/ or plant services depend upon [5, 6].

- the quality of the source of makeup water;
- the manner in which the water is used;
- environmental regulations; and
- Site climatic conditions governing wastewater disposal.

In wastewater treatment systems, suspended solids can be removed by physical treatment to some extent. The removal of biodegradable organics, suspended solids and pathogens is achieved through the secondary treatment operation units. The more stringent standards are dealt with the removal of nutrients and priority pollutants.

When wastewater is to be reused, regulations normally include requirements for the removal of refractory organics, heavy metals and, in some cases, dissolved inorganic solids [7–10].

Process waters and all other special drainages throughout the plant/refinery shall be isolated from surface runoff. The surface drainage shall be collected in a dedicated and separate clean stormwater sewer system. Extensive efforts shall be made to segregate the surface drainages and to avoid the contamination or mixing with the oily water sewers.

In oil, gas and chemical industries, water is the most commonly used agent for controlling and fighting a fire, by cooling adjacent equipment, and for controlling and/or extinguishing the fire either by itself or combined as a foam. It can also provide protection for firefighters and other personnel in the event of fire. Water shall therefore be readily available at all the appropriate locations, at the correct pressure and in the required quantity [7–10].

These factors should be considered in selecting the overall plant process and utility systems.

1.2 Quality of Source Waters

The quality of many water sources will change little over the lifetime of treatment plant except for the seasonal changes that should be anticipated in advance. In some instances, it is best arrived at by judgment based on past trends in quality, a survey of the source and evaluation of future developments relating to the supply. Other sources can be expected to deteriorate substantially as a result of an increase in wastes. A reasonably accurate prediction of such changes in quality is difficult to make.

Groundwater sources tend to be uniform in quality, to contain greater amount of dissolved substances, to be free of turbidity and to be low in color. Surface water supplies receive greater exposure to wastes, including accidental spills of the variety of substances [11–13].

Silt may enter in the water supply. Depending on the source of the supplied water to the plant/refinery and the characteristics and impurities, provision of sedimentation before the water is used shall be investigated. Special attention shall be made to reduce deposition of solids in cooling tower basins, heat exchangers and other consumers and also to prevent these solids from entering oily water sewers [12–15].

In chemical plants, water formed by chemical reaction is generally less than the water evaporated into atmosphere, so that the water discharged tends to be less than the water intake. However, most part of the intake water is discharged. Also, rain water is fouled while flowing through contaminated areas of a plant and is discharged as part of the wastewater [18–20].

Generalizations like the above, although useful, are not a substitute for the definitive information required for plant design.

To provide adequate protection against pollution, special studies in the design of intakes should have to be made to indicate the most favorable locations for obtaining water.

In connection with deep reservoirs, multiple intakes offer flexibility in selecting water from various depths, thus overcoming poorer water quality resulting from seasonal changes. For groundwater sources, the location and depths of wells should be considered in order to avoid pollution and secure water of favorable quality.

The water contaminant parameters determined in oil, gas and chemical processing industries include biologic oxygen demand (BOD_5), chemical oxygen demand (COD), oil, total suspended solids (TSS), ammonia (NH_3), phenolics, hydrogen sulfide (H_2S), trace organics and some heavy metals. Table 1.1 shows the major sources of each of the contaminants [5–8].

Table 1.1 Water and wastewater pollutions and sources

Pollution	Sources
BOD_5, COD, Oil	Process wastewater, cooling tower blowdown (if hydrocarbons leak into cooling water system), ballast water, tank flow drainage and runoff
Total suspended solids	Process wastewater, cooling tower blowdown, ballast water, tank flow drainage and runoff
Phenolics	Process wastewater (particularly from fluid catalytic cracking unit)
NH_3, H_2S, trace organics	Process wastewater (particularly from fluid catalytic cracking unit and coker)
Heavy metals	Process wastewater, tankage wastewater discharges, cooling tower blowdown (if chromate-type cooling water treatment chemicals are used)

1.3 Source Water Types

It is clear that one of the criteria for location of the water supplies will be the availability of water, but it is also a good practice to set up a preliminary layout of the water supply system in order to ensure an even utilization of the forage resources and plan the additional investigations necessary for providing the water supplies [5–8]. Source or makeup water is normally either groundwater or surface water, neither of which is ever chemically pure.

Groundwaters contain dissolved inorganic impurities that come from the rock and sand strata through which the water is passed. Surface waters often contain silt particles in suspension (suspended solids) and dissolved organic impurities (dissolved solids). Table 1.2 lists some of the common properties or characteristics and the normal constituents of water, together with corresponding associated operating difficulties and potential methods of water treatment [5–8].

1.4 Preliminary Water Treatment

Preliminary treatment or pre-treatment is any physical, chemical or mechanical process used on water before it undergoes the main treatment process. During preliminary treatment, screens may be used to remove rocks, sticks, leaves and other debris; chemicals may be added to control algae growth; and a pre-sedimentation stage can settle out sand, grit and gravel from raw water [25–30].

Regardless of the final use of source water and any subsequent treatment, it is often advisable to carry out this general treatment close to the intake or well. The purpose of treatment close to source is to protect the distribution system itself and at the same time to provide initial or sufficient treatment for some of the main uses of water.

In case of surface water, general protection should be provided against clogging and deposits. The obstruction or clogging of apertures and pipes by foreign matter can be avoided by screening or straining through a suitable mesh. The protection used is either a bar screen, in which the gap between the bars can be as narrow as 2 mm, or a drum or belt filter, with a mesh of over 250 micrometers (μm) [26–31].

According to the requirements of the equipment and the amount of pollution (slime) in the water, a 250-μm filter may be used on an open system, or with microstraining, a 50 μm may be necessary. In some cases, rapid filtration through siliceous sand or diatomaceous earth may be necessary after screening, which will eliminate suspended matter down to a few micrometers. Where there is large amount of suspended matter, grit removal and/or some degree of settling should be provided [27–33].

In case of groundwater, the main risks are abrasion by sand or corrosion. For abrasion, the pumps should be suitably designed, and the protection, which concerns only the parts of the system downstream the pumps, will take the form of very

Table 1.2 Common impurities and characteristics in water [9–39]

Constituent	Chemical formula	Difficulties caused	Means of treatment
Hardness	Calcium, magnesium, barium and strontium salts expressed as $CaCO_3$	Chief source of scale in heat exchange equipment, boilers, pipelines, etc.; forms curds with soap; interferes with dyeing, etc	Softening, distillation, internal boiler water treatment, surface-active agents, reverse osmosis, electrodialysis
Turbidity	None, usually expressed in Jackson turbidity units	Imparts unsightly appearance to water; deposits in water lines, process equipment, boilers, etc.; interferes with most process uses	Coagulation, settling and filtration
Alkalinity	Bicarbonate HCO_3^-, carbonate (CO_3^{2-}) and hydroxyl (OH^-) expressed as $CaCO_3$	Foaming and carryover of solids with steam; embrittlement of boilers steel; bicarbonate and carbonate produce CO_2 in steam, a source of corrosion	Lime and lime-soda softening acid treatment, hydrogen zeolite softening, demineralization, dealkalization by anion exchange, distillation, degasify
Color	None	Decaying organic material and metallic ions causing color may cause foaming in boilers; hinders precipitation methods, such as iron removal, hot phosphate softening, can stain product in process use	Coagulation, filtration, chlorination, adsorption by activated carbon
Carbon dioxide	CO_2	Corrosion in water lines and particularly steam and condensate lines	Aeration, deaeration, neutralization with alkalies, filming and neutralizing amines
PH	Hydrogen ion concentration defined as: $pH = \log\left(\frac{1}{H^+}\right)$	pH varies according to acidic or alkaline solids in water; most natural waters have a pH of 6.0–8.0	pH can be increased by alkalies and decreased by acids
Sulfate	SO_4^{2-}	Adds to solids' content of water, but in itself is not usually significant; combines with calcium to form calcium sulfate scale	Demineralization, distillation, reverse osmosis, electrodialysis

(continued)

Table 1.2 (continued)

Constituent	Chemical formula	Difficulties caused	Means of treatment
Chloride	Cl^{-1}	Adds to solids' content and increases corrosive character of water	Demineralization, distillation, reverse osmosis, electrodialysis
Nitrate	NO_3^-	Adds to solids' content, but is not usually significant industrially; useful for control of boiler metal embrittlement	Demineralization, distillation, reverse osmosis
Fluoride	F^{-1}	Not usually significant industrially	Adsorption with magnesium hydroxide, calcium phosphate, bone black, alum coagulation, reverse osmosis, electrodialysis
Silica	SiO_2	Scale in boilers and cooling water systems; insoluble turbine blade deposits due to silica vaporization	Hot process removal with magnesium salt adsorption by highly basic anion exchange resins, in conjunction with demineralization, distillation
Iron	Fe^{2+} (ferrous) Fe^{3+} (ferric)	Discolors water on precipitation; source of deposits in water lines, boilers, etc.; interferes with dyeing, tanning paper manufacturer, etc.	Aeration, coagulation and filtration, lime softening, cation exchange, contact filtration, surface-active agents for ion retention
Manganese	Mn^{2+}	Same as iron	Same as iron
Oil	Expressed as oil or chloroform extractable matter, ppm	Scale, sludge and foaming in boilers; impedes heat exchange; undesirable in most processes	Baffle separators, strainers coagulation and filtration, diatomaceous earth filtration
Oxygen	O_2	Corrosion of water lines, heat exchange equipment, boilers, return lines, etc.	Deaeration, sodium sulfite, corrosion inhibitors, hydrazine or suitable substitutes
Hydrogen sulfide	H_2S	Cause of "rotten egg" odor; corrosion	Aeration, chlonnation, highly basic anion exchange
Ammonia	NH_3	Corrosion of copper and zinc alloys by formation of complex soluble ion	Cation exchange with hydrogen zeolite, chlorination, deaeration, mixed-bed demineralization

(continued)

Table 1.2   (continued)

Constituent	Chemical formula	Difficulties caused	Means of treatment
Conductivity	Expressed as micromhos-specific conductance	Conductivity is the result of ionizable solids in solution; high conductivity can increase the corrosive characteristics of a water	Any process which decreases dissolved solids' content will decrease conductivity; examples are demineralization, lime softening
Dissolved solids	None	"Dissolved solids" is measure of total amount of dissolved matter, determined by evaporation; high concentrations of dissolved solids are objectionable because of process interference and as a cause of foaming in boilers	Various softening process, such as lime softening and cation exchange by hydrogen zeolite, will reduce dissolved solids, demineralization, distillation, reverse osmosis, electrodialysis
Suspended solids	None	"Suspended solids" is the measure of undissolved matter, determined gravimetrically; suspended solid plug lines cause deposits in heat exchange equipment, boilers, etc.	Subsidence, filtration, usually preceded by coagulation and settling
Total solids	None	"Total solids" is the sum of dissolved and suspended solids, determined gravimetrically	See "Dissolved solids" and "Suspended solids"

rapid filtration through sand, straining under pressure of use of hydrocyclones, if the grit is of the right grain size. Moreover, corrosion frequently occurs on systems carrying underground water and leads to the formation of tuberculi and concretions. This corrosivity is often caused by the lack of oxygen [35–39].

The best method, therefore, of preventing corrosion is by oxygenation and by filtration processes that have the dual advantage of removing the grit and any iron present and of feeding into the water the minimum amount of oxygen needed for system to protect itself [40–43]. Figure 1.1 shows potable water production from a river.

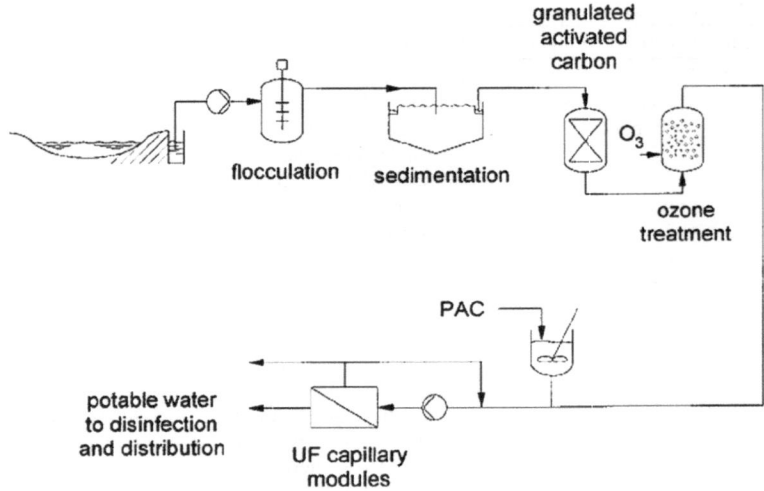

Fig. 1.1 Potable water production from a river (Rautenbach and Voßenkaul [51] © Elsevier, 2001, reprinted with permission)

1.4.1 Dissolved Oxygen Saturation Concentrations in Aquatic Systems

In this section, a simple predictive tool for dissolved oxygen saturation concentrations in aquatic systems as a function of chloride concentration and temperature using a novel Arrhenius-type asymptotic exponential function that was formulated by Bahadori and Vuthaluru [20] is presented.

This model predicts the dissolved oxygen saturation concentrations for temperatures up to 50 °C and chloride concentrations up to 25 gram per liter. Estimations are found to be in excellent agreement with the reliable data in the literature with average absolute deviation being 3 %.

The tool developed by Bahadori and Vuthaluru [20] can be of immense practical value for the engineers and scientists to provide a quick check on the oxygen saturation concentrations in aquatic systems without opting for any experimental measurements.

In particular, environmental science experts may find the approach to be user-friendly with transparent calculations involving no complex expressions.

Equation 1.1 represents the proposed governing equation in which four coefficients are used to correlate the oxygen saturation concentrations in aquatic systems as a function of temperature and chloride concentration where the relevant coefficients have been reported in Table 1.3.

$$\ln(C_O) = a + \frac{b}{T} + \frac{c}{T^2} + \frac{d}{T^3} \tag{1.1}$$

Coefficient	Fitted parameter values
A_1	$-2.90214864273 \times 10^1$
B_1	$-3.57039727398 \times 10^{-1}$
C_1	$5.603855782563 \times 10^{-2}$
D_1	$-1.99736345182 \times 10^{-3}$
A_2	$2.60884743643 \times 10^4$
B_2	$3.08829824442 \times 10^2$
C_2	$-5.00051320536 \times 10^1$
D_2	1.788331812403
A_3	$-7.7438517753 \times 10^6$
B_3	$-8.96406240764 \times 10^4$
C_3	$1.48584285199 \times 10^4$
D_3	$-5.33193086284 \times 10^2$
A_4	8.1489174667×10^8
B_4	$8.46725959984 \times 10^6$
C_4	$-1.47047175489 \times 10^6$
D_4	5.2945414173×10^4

Table 1.3 Tuned coefficients used in Eqs. 1.2–1.5

where

$$a = A_1 + B_1\psi + C_1\psi^2 + D_1\psi^3 \qquad (1.2)$$

$$b = A_2 + B_2\psi + C_2\psi^2 + D_2\psi^3 \qquad (1.3)$$

$$c = A_3 + B_3\psi + C_3\psi^2 + D_3\psi^3 \qquad (1.4)$$

$$d = A_4 + B_4\psi + C_4\psi^2 + D_4\psi^3 \qquad (1.5)$$

These optimum-tuned coefficients help to cover the oxygen saturation concentrations in aquatic systems (C_o) for temperatures up to 50 °C as well as chloride concentrations up to 25 g/L. The optimum-tuned coefficients given in Table 1.3 can be retuned quickly according to proposed approach if more data are available in the future.

Below is the list of symbol used in the Eqs. 1.1–1.5:

A Tuned coefficient; B Tuned coefficient; C Tuned coefficient; D Tuned coefficient; o dissolved oxygen saturation concentrations in aquatic systems, mg/L; T Temperature; K; ψ Chloride concentration; g/L.

The proposed novel tool in the present work is simple and unique expression which is non-existent in the literature. Furthermore, we have selected exponential function to develop the tool which has well-behaved (i.e., smooth and non-oscillatory) equations enabling fast and more accurate predictions.

Figure 1.2 shows the predicted results from the proposed predictive tool for the oxygen saturation concentrations in aquatic systems (C_o) as a function of and temperature with the reported data in the literature [22].

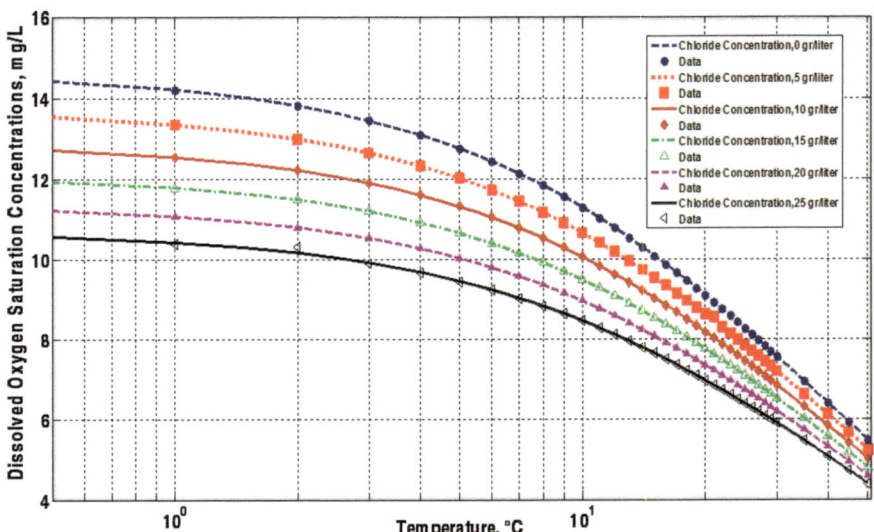

Fig. 1.2 Oxygen saturation concentrations in aquatic systems as a function of chloride concentration and temperature with the reported data (Bahadori and Vuthaluru [22] © Elsevier, 2010, reprinted with permission)

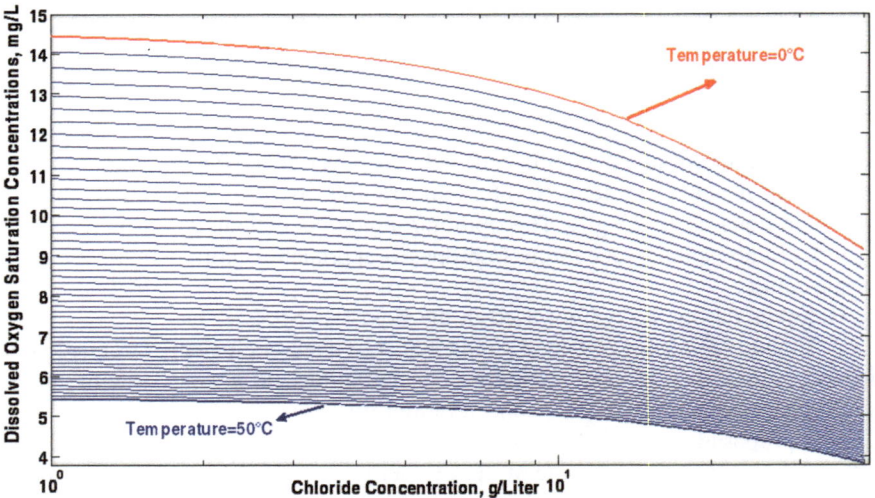

Fig. 1.3 Performance of proposed predictive tool for the prediction of the oxygen saturation concentrations in aquatic systems as a function of aqueous chloride concentration and temperature (Bahadori and Vuthaluru [22] © Elsevier, 2010, reprinted with permission)

It is evident from the figure that there is a good agreement between predicted values [for wide range of chloride concentrations (ψ) and temperatures] and the reliable data [34, 48].

Figure 1.2 shows that the oxygen saturation concentrations in aquatic systems decrease at high temperature and high chloride concentrations. Figure 1.3 shows the performance of proposed correlation for wide range of conditions.

1.5 Treatment Process Selection

The quality of the source, giving due consideration to variations and possible future changes, the quality goals for the finished water and cost, shall form the basis for selecting a treatment process [43–48].

Often, various types and combination of treatment units would be used to achieve the performance desired (see Table 1.2). Determination of the most suitable plan should be on a comparative cost study which includes an evaluation of the merits and liabilities of each proposal.

The experience acquired through the treatment of the same or similar source shall provide an excellent guide in selecting a plan.

Where experience is lacking or where there is the desire for a different degree of performance, special studies should be indicated. Tests conducted in the laboratory, in existing plants or in pilot plants should then be employed to obtain information for design purposes [40–49].

1.6 Potable Water Quality

In oil, gas and chemical processing refineries and/or plants' water system, the minimum specification required for potable water should be as per local authority of municipal water supply and is recommended to be based on establishment water system performance goals potable water standard specifications as required by relevant national standards.

Table 1.4 Comparison of physical characteristics of water

Characteristic	Recommended limits[a]	AWWA goals (the figures in this column are taken from AWWA Committee Report)
Turbidity units	5	<0.10
Color units	15	<3
Odor (threshold odor number)	3	No odor
Taste	–	None objectionable

[a] Limits that should not be exceeded according to latest USPHS Drinking Water Standards

1.6.1 Physical Characteristics

Table 1.4 compares the levels for turbidity, color, taste and odor [7].

The committee report presents those levels that should be approached by well-designed and operated systems and which reflect a high degree of consumer acceptability.

1.6.2 Chemical Characteristics

Table 1.5 gives the maximum concentration of various chemical substances allowed by the USPHS standards and recommended by the AWWA Report [7], and the recommended lower, optimum and upper control limits for fluoride concentrations are shown in Table 1.6.

The AWWA Committee Report establishes a hardness of 80 mg/L (as $CaCO_3$) as the desirable objective for potable water supplies. Although this is not of

Table 1.5 Comparison of chemical characteristics

Compound	Concentration that should not be exceeded (mg/L)	Concentration which, if exceeded, constitutes grounds for rejection of supply (mg/L)
Alkylbenzene sulfonate (ABS)	0.5	–
Arsenic (As)	0.01	0.05
Barium (Ba)		1
Chloride (Cl)	250	–
Cadmium (Cd)	–	0.01
Chromium		0.05
Copper (Cu)	1	–
Carbon chloroform extract (CCE)	0.2	–
Cyanide (CN)	0.01	0.2
Iron (Fe)	0.30	
Lead (Pb)	–	0.05
Manganese (Mn)	0.05	–
Nitrate (NO_3)	45	–
Phenols	0.001	–
Selenium (Se)	–	0.01
Silver (Ag)	–	0.05
Sulfate (SO_4)	250	–
Total dissolved solids	500	–
Zinc (Zn)	5	–

Table 1.6 Allowable fluoride concentration

Annual average of maximum daily air temperature, based on temperature data obtained for a minimum of 5 years	Lower recommended control limits, fluoride concentrations (mg/L) (from "Drinking Water Standards," US Public Health Service, No. 956, 1962)	Optimum recommended control limits, fluoride concentrations (mg/L) (from "Drinking Water Standards," US Public Health Service, No. 956, 1962)	Upper recommended control limits, fluoride concentrations (mg/L) (from "Drinking Water Standards," US Public Health Service, No. 956, 1962)
12 or lower	0.9	1.2	1.7
12.1–14.6	0.8	1.1	1.5
14.6–17.7	0.8	1	1.3
17.7–21.4	0.7	0.9	1.2
21.5–26.2	0.7	0.8	1
26.3–32.5	0.6	0.7	0.8

concern to utilities with sources of supply that are naturally quite soft, it is significant to utilities in hard water areas because softening must be provided to achieve this goal.

A number of limits expressed in Table 1.6 are based on aesthetic rather than on health considerations. For example, the limits for iron and manganese are based on the staining and other objectional properties of these elements.

Fluoride is considered as an essential constituent of drinking water for prevention of tooth decay in children. Conversely, excess fluoride may give rise to dental fluorosis (spotting of the teeth) in children. In the Drinking Water Standards is also recommended that fluoride in average concentrations greater than twice the optimum values shall constitute grounds for rejection of the supply (see Table 1.6).

1.6.3 Radioactivity

In establishing reasonable, long-term limits for radioactivity in potable water, the population's total exposure to radiation must be considered. This requires the assessment of the intake from sources such as food and milk, as well as from potable water, together with an evaluation of the effects of specific radioactive substances.

The USPHS standards specify that water supplies other sources of radioactivity intake of radium-226 and strontium-90 when the water contains these substances in amounts not exceeding 3 pico curie (pCi) per liter and 10 pCi (10×10^{-12} Ci) per liter, respectively [7].

In the known absence of strontium-90 and alpha emitters, the water supply is acceptable when the known concentration does not exceed 1,000 pCi per liter.

A thorough considerable knowledge has been acquired on the treatment of water to remove radioactive substances; it is desirable to maintain the levels of radioactivity in raw water well below the established limits.

1.7 Boiler Water Quality Criteria

There are four types of impurities of concern in water to be used for the generation of steam [5]:

1. Scale-forming solids which are usually the salts of calcium and magnesium along with boiler corrosion products. Silica, manganese and iron can also form scale.
2. The much more soluble sodium salts which do not normally form scale, but can concentrate under scale deposits to enhance corrosion or in the boiler water to increase carryover due to boiler water foaming.
3. Dissolved gases, such as oxygen and carbon dioxide, which can cause corrosion.
4. Silica which can volatilize with the steam in sufficient concentrations to deposit in steam turbines.

Blowdown should be employed to maintain boiler water–dissolved solids at an appropriate level of concentration. At equilibrium, the quantities of dissolved solids removed by blowdown exactly equal those introduced with the feedwater plus any injected chemicals [5].

1.7.1 Sludge and Total Suspended Solids

These result from the precipitation in the boiler of feedwater hardness constituents due to heat and due to interaction between treatment chemicals and from corrosion products in the feedwater. They can contribute to boiler tube deposits and enhance foaming characteristics, leading to increased carryover [5].

1.7.2 Total Dissolved Solids

These consist of all salts naturally present in the feedwater of soluble silica and of any chemical treatment added. Dissolved solids do not normally contribute to scale formation, but excessively high concentrations can cause foaming and carryover or can enhance "under deposit" boiler tube corrosion.

1.7.3 Silica

This may be the blowdown controlling factor in softened water containing high silica. High boiler water silica content can result in silica vaporization with the steam and, under certain circumstances, siliceous scale. This is illustrated by silica solubility. Silica content of boiler water is not as critical for steam systems without steam turbines [5–21].

Silica content of the boiler water is critical for steam turbines and scaling of boiler heat transfer surfaces. Silica (SiO_2) can volatilize with the steam in sufficient concentrations to deposit in steam turbines, leading to scale formation on boiler surfaces. In this work, a simple correlation is presented to predict silica (SiO_2) solubility in steam of boilers as a function of pressure and water silica content. The solubility of silica in steam directly depends on both the density and the temperature of steam. With decreasing temperature and density, solubility of silica reduces. As the pressure affects steam density which has a strong bearing on steam temperature, it has an important effect on the solubility of silica in steam [21].

The proposed correlation predicts the solubility of silica (SiO_2) in steam for pressure up to 22,000 kPa and boiler water silica contents up to 500 mg/kg. The predictions from the proposed correlation have been compared with reported data and found good agreement with average absolute deviation being around 4 %. This simple-to-use correlation can be of immense practical value for the engineers to have a quick check on silica (SiO_2) solubility in steam of boilers as a function of pressure and water silica content at various conditions without performing any experimental measurements. In particular, personnel dealing with the utility boilers would find the proposed approach to be user-friendly involving no complex expressions with transparent calculations [21].

Equation 1.6 represents the proposed governing equation in which four coefficients are used to correlate the solubility of silica in steam of boilers as a function of pressure and silica water content where the relevant coefficients have been reported in Table 1.7.

$$\ln(S) = a + bP + cP^2 + dP^3 \qquad (1.6)$$

where

$$a = A_1 + B_1(C_W) + C_1\left(C_W^2\right) + D_1\left(C_W^3\right) \qquad (1.7)$$

$$b = A_2 + B_2(C_W) + C_2\left(C_W^2\right) + D_2\left(C_W^3\right) \qquad (1.8)$$

$$c = A_3 + B_3(C_W) + C_3\left(C_W^2\right) + D_3\left(C_W^3\right) \qquad (1.9)$$

$$d = A_4 + B_4(C_W) + C_4\left(C_W^2\right) + D_4\left(C_W^3\right) \qquad (1.10)$$

Table 1.7 Tuned coefficients used in Eqs. 1.7–1.10 [21]

Coefficient	Coefficients for boiler water silica less than 50 mg/kg	Coefficients for boiler water silica between 50 and 500 mg/kg
A_1	-9.91036274604	-7.29715827822
B_1	$3.6035324119 \times 10^{-1}$	$3.36188512456 \times 10^{-2}$
C_1	$-1.30855233907 \times 10^{-2}$	$-1.20354059517 \times 10^{-4}$
D_1	$1.50175517061 \times 10^{-4}$	$1.29245662436 \times 10^{-7}$
A_2	$4.09783293899 \times 10^{-4}$	$4.18453241026 \times 10^{-4}$
B_2	$1.57322016793 \times 10^{-5}$	$5.55318691455 \times 10^{-8}$
C_2	$-1.0136656004 \times 10^{-6}$	$-1.7039052649 \times 10^{-9}$
D_2	$1.3853045954 \times 10^{-8}$	$5.40562939438 \times 10^{-12}$
A_3	$2.37599942152 \times 10^{-9}$	$2.78159673634 \times 10^{-9}$
B_3	$-1.20699785241 \times 10^{-9}$	$1.098784264 \times 10^{-11}$
C_3	$7.68541076834 \times 10^{-11}$	$5.52027970387 \times 10^{-15}$
D_3	$-1.01195964275 \times 10^{-12}$	$-2.51031446555 \times 10^{-16}$
A_4	$-4.79364161851 \times 10^{-14}$	$-8.17515244627 \times 10^{-14}$
B_4	$2.79052818094 \times 10^{-14}$	$-6.16067430892 \times 10^{-16}$
C_4	$-1.77452718145 \times 10^{-15}$	$2.81610902151 \times 10^{-18}$
D_4	$2.2667148187 \times 10^{-17}$	$2.010057934 \times 10^{-21}$

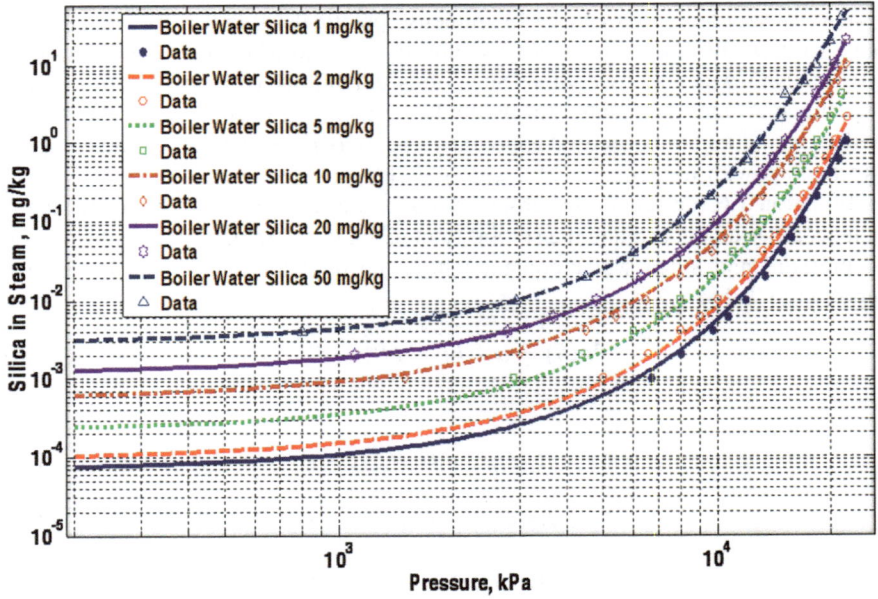

Fig. 1.4 Comparison of predicted solubility of silica against literature-reported data for low concentration of boiler water silica (1–50 mg/kg) (Bahadori and Vuthaluru [21] © Elsevier, 2010, reprinted with permission)

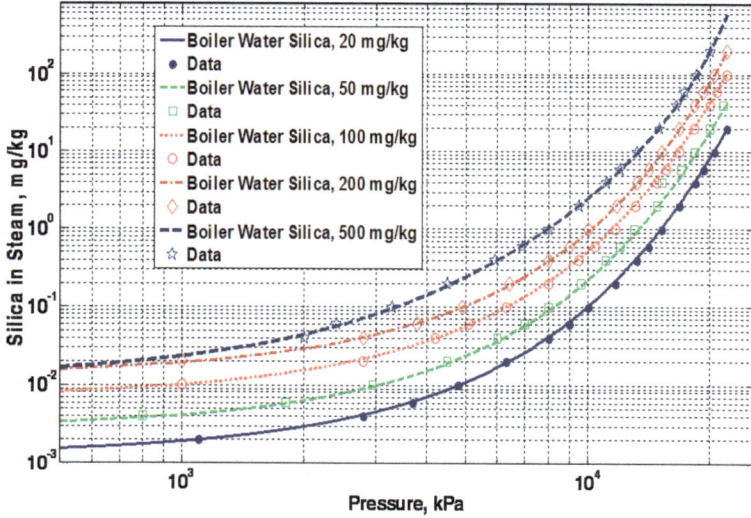

Fig. 1.5 Comparison of predicted solubility of silica against literature-reported data [21] for high concentration of boiler water silica (20–500 mg/kg). (Bahadori and Vuthaluru [21] © Elsevier, 2010, reprinted with permission)

In the above equations, *A*, *B*, *C* and *D* are tuned coefficients, *P* is pressure in kPa, C_w is boiler water silica in mg/kg, and *S* is silica concentration in steam in mg/kg.

These optimum-tuned coefficients help to cover the solubility of silica in steam of boilers for pressures up to 22,000 kPa(g) as well as boiler water up to 500 mg/kg.

Figures 1.4 and 1.5 show the results of the proposed correlation for predicting the silica solubility in steam of boilers as a function of pressure and water silica content in comparison with the reported data. It is evident from the above figures that there is a good agreement between predicted values (for wide range of pressure and water silica contents) and the reported data in literature.

The solubility of silica in steam directly depends on both the density and the temperature of steam. With decreasing temperature and density, solubility of silica reduces. As the pressure affects steam density which has a strong bearing on steam temperature, it has an important effect on the solubility of silica in steam. These figures show that the solubility of silica in steam increases at high pressures and high boiler water silica [21].

If the proposed approach is adopted in utilities on a periodic basis, significant savings can be assured with reduced maintenance issues in terms of failure of boiler surfaces and scaling problems associated with the carryover of silica in steam of boilers. Current efforts in this investigation pave the way for alleviating the problems associated with the overheating and failure of boiler sections due to scale formation and turbine inefficiencies by arriving at an accurate measure of silica solubility in steam which can be used by the utility personnel for monitoring the operational parameters.

1.7.3.1 Sample Calculation for the Practice Engineers

The calculations shown here are for a typical boiler operating with 500 mg/kg silica in the water at 2,760 kPa. An estimation of silica which comes out of the system when the steam is expanded through a turbine to 690 kPa is given below.

Solution

As the boiler water silica is more than 50 mg/kg, second column of Table 1.7 is used:

$a = -4.4205397$ (from Eq. 1.7)
$b = 6.9594653 \times 10^{-4}$ (from Eq. 1.8)
$c = -2.1723342 \times 10^{-8}$ (from Eq. 1.9)
$d = 5.6549925 \times 10^{-13}$ (from Eq. 1.10)
$S = 0.0746$ mg/kg (from Eq. 1.6).

When the steam is expanded through a turbine to 690 kPa. Because "a," "b," "c" and "d" are as a function of boiler water silica so we have the same coefficients, however, for new pressure ($P = 690$ kPa) we will have

$S = 0.0205$ mg/kg (from Eq. 1.6)

So a 2,760 kPa (ga) boiler operating with 500 mg/kg SiO_2 in the water within the boiler could generate steam containing 0.0746 mg/kg of SiO_2. When this steam is expanded through a turbine to 690 kPa (ga), the solubility of SiO_2 decreases to about 0.0205 mg/kg. The silica coming out of solution ($0.0746 - 0.0205 = 0.054$ mg/kg) could coat turbine blades and eventually result in extensive turbine maintenance.

This is classic example showing how the information evolving out of this correlation can be used to understand and predict the potential maintenance issues which could damage the utility components. The results of this example

Table 1.8 Recommended boiler water limits and associated steam purity at steady-state, full-load operation, drum-type boilers [5, 6]

Drum pressure bar (ga)	Range of total dissolved solids in boiler water mg/kg (max)	Range of total alkalinity mg/kg (max)	Suspended solids in boiler water mg/kg (max)	Range of total dissolved solids in steam mg/kg (max expected valued)
0.20–69	700–3,500	140–700	15	0.2–1.0
20.76–31.03	600–3,000	120–600	10	0.2–1.0
31.10–41.38	500–2,500	100–500	8	0.2–1.0
41.45–51.72	400–2,000	80–400	6	0.2–1.0
51.79–62.07	300–1,500	60–300	4	0.2–1.0
62.14–68.96	250–1,250	50–250	2	0.2–1.0

Table 1.9 Common deposits formed in water systems

Name	Chemical composition	Deposit formation at T < 100 °C with or without evaporation	Deposit formation at T > 100 °C with no evaporation	Evaporation at T > 100 °C	Water vapor or steam
Acmite	$Na_2O.Fe_2O_3.4SiO_2$			✔	
Analctive	$Na_2O.Al_2O_3.4SiO_2.2H_2O$			✔	✔
Anhydrite	$CaSO_4$		✔	✔	
Aragonite	$CaCO_3$	✔	✔	✔	
Biologic—non-spore bacteria		✔			
Biologic—spore bacteria		✔			
Biologic—Fungi		✔			
Biologic—Algae and diatoms		✔			
Biologic—Crustaceans		✔			
Brucite	$Mg(OH)_2$		✔	✔	
Burkeite	$Na_2CO_3.2Na_2SO_4$				✔
Calcite	$CaCO_3$	✔	✔	✔	
Calcium hydroxide	$Ca(OH)_2$			✔	
Carbonaceous		✔	✔	✔	✔
Copper	Cu			✔	
Cuprite	Cu_2O		✔		
Ferrous oxide	FeO		✔		
Goethite	$Fe_2O_3.H_2O$	✔	✔	✔	
Gypsum	$CaSO_4.2H_2O$	✔	✔	✔	
Halite	$NaCl$				✔
Hydroxyapatite	$Ca_{10}(PO_4)_6(OH)_2$ phosphate(basic)	✔	✔	✔	
Magnesium	$Mg_3(PO_4)_2.Mg(OH)_2$		✔	✔	
Magnetite	Fe_3O_4		✔	✔	✔
Oil (chloroform extractable)		✔	✔	✔	✔
Quartz	SiO_2				✔
Serpentine	$3MgO.2SiO_2.2H_2O$		✔	✔	
Siderite	$FeCO_3$				✔
Silica (amorphous)	SiO_2				✔
Sodium carbonate	Na_2CO_3				✔

(continued)

Table 1.9 (continued)

Name	Chemical composition	Deposit formation at T $<$ 100 °C with or without evaporation	Deposit formation at T $>$ 100 °C with no evaporation	Evaporation at T $>$ 100 °C	Water vapor or steam
Sodium disilicate	$Na_2Si_2O_6$				✔
Sodium ferrous phosphate	$NaFePO_4$			✔	
Sodium silicate	Na_2SiO_3				✔
Tenorite	CuO			✔	
Thenardite	Na_2SO_4			✔	✔
Xonotlite	$5CaO.5SiO_2.H_2O$			✔	

(0.054 mg/kg) have good agreement with calculated results based on literature-based data of 0.06 mg/kg.

Iron

Occasionally in high pressure boilers where the iron content is high in relation to total solids, blowdown may be based upon controlling iron concentrations; high concentrations of suspended iron in boiler water can produce serious boiler deposit problems and are often indications of potentially serious corrosion in the steam/ steam condensate systems.

While there are other considerations (such as corrosive- or deposit-forming tendencies) in establishing limits for boiler water composition, Table 1.8 clearly indicates that boiler feedwater purity becomes more important as operating pressures [5].

1.8 Common Deposits Formed in Water Systems

The deposits may be classified generally as scale, sludge, corrosion products and biologic deposits. The most common types of deposits are shown in Table 1.9: Symbol "✔" means the formation of deposits in the relevant conditions.

Chapter 2
Processes Design

Keywords Coagulation · Flocculation · Sedimentation · Clarification system · Filter · Ion exchange · Demineralizing unit · Reverse osmosis · Electrodialysis

Industrial raw water treatment is used to optimize most water-based industrial processes, such as heating, cooling, processing, cleaning and rinsing processes. The ultimate goal is to reduce operating costs and risks. Poor water treatment can cause serious damage to the process and the final results. Surfaces of pipes and vessels can be affected by corrosion, and steam boilers can scale up or corrode [52].

The suspended particles in water vary considerably in source, composition, charge, particle size, shape and density. The smaller particles present in water are kept in suspension by the action of physical forces on the particles themselves. Figure 2.1 shows an overview of the water treatment process:

The water treatment process consists of several steps:

Coagulation and flocculation

Coagulation and flocculation processes are used to separate the suspended solids (SS) portion from the water. Raw water from terminal reservoirs is drawn into mixing basins at our treatment plants where we add alum, polymer and sometimes lime and carbon dioxide. This process causes small particles to stick to one another, forming larger particles [53–56].

Coagulation and flocculation occur in successive steps intended to overcome the forces stabilizing the suspended particles, allowing particle collision and growth of floc. The purpose of coagulation and flocculation is to cause small pollutant particles such as metals to aggregate and form large enough floc, so that they can be separated from the wastewater through sedimentation.

There are three main types of coagulants that are used to overcome the repulsive forces of particles, thus causing them to aggregate. Electrolytes, organic polymers and synthetic polyelectrolytes are added to wastewater, and then

A. Bahadori et al., *Essentials of Water Systems Design in the Oil, Gas, and Chemical Processing Industries*, SpringerBriefs in Applied Sciences and Technology, DOI: 10.1007/978-1-4614-6516-4_2, © The Author(s) 2013

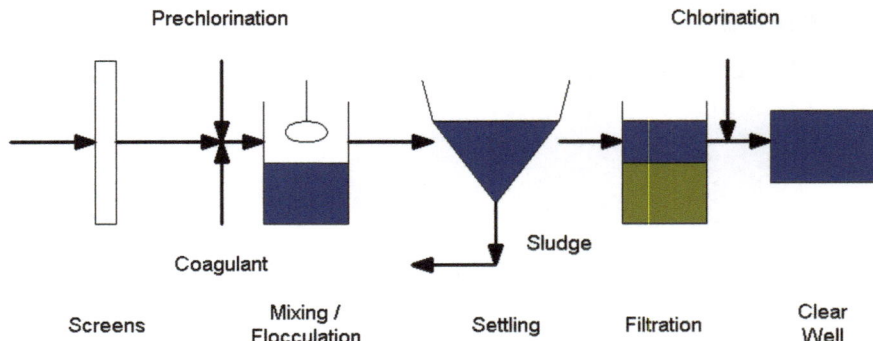

Fig. 2.1 An overview on water treatment process

flocculation tanks mix the water to promote flocs and subsequent physical separation [57–59].

Each of these processes is briefly explained in below Figs. 2.2 and 2.3:

Sedimentation

Over time, the now-larger particles become heavy enough to settle to the bottom of a basin from which sediment is removed.

Filtration

The water is then filtered through layers of fine, granulated materials—either sand or sand and coal, depending on the treatment plant. As smaller, suspended particles are removed, turbidity diminishes and clear water emerges [59–61].

Disinfection

To protect against any bacteria, viruses and other microbes that might remain, disinfectant is added before the water flows into underground reservoirs throughout the distribution system and into your home or business. Denver Water carefully monitors the amount of disinfectant added to maintain quality of the water at the farthest reaches of the system. Fluoride occurs naturally in our water but also is added to treated water [61–63].

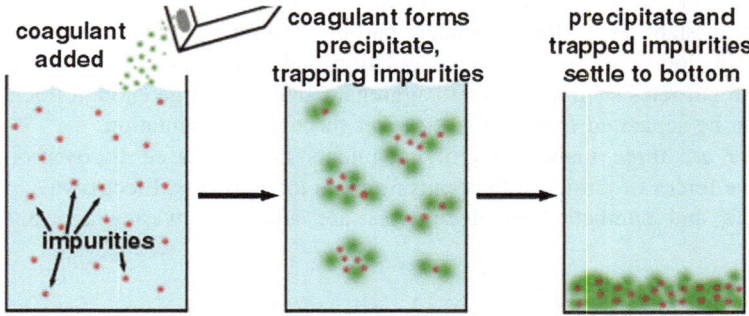

Fig. 2.2 Schematic of coagulation process

Fig. 2.3 Schematic of
flocculation process

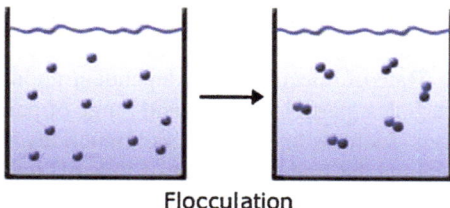

Flocculation

Corrosion control

pH is maintained by adding alkaline substances to reduce corrosion in the
distribution system and the plumbing in home or business.

2.1 Coagulation and Flocculation

2.1.1 General Information

In modern water treatment, coagulation and flocculation are still essential components of the overall suite of treatment processes. Coagulation is always considered along with flocculation and is used to remove particles which cannot be
removed by sedimentation or filtration alone. These particles are usually less than
1 μm in size and are termed colloids. They have poor settling characteristics and
are responsible for the color and turbidity of water. They include clays, metal
oxides, proteins, microorganisms and organic substances such as those that give
the brown coloration to water from "peaty" catchment areas. The important
property which they all have is that they carry a negative charge and this, along
with the interaction between the colloidal particles and the water, prevents them
from aggregating and settling in still water. The particles can be aggregated by
adding either multivalent ions or colloids having an opposite (positive) charge.
These are added as chemical coagulants [61–64].

Chemicals commonly used as coagulants in water treatment are aluminum and
ferric salts which are present as the ions Al^{3+} and Fe^{3+}. These positively charged
multivalent ions neutralize the naturally occurring negatively charged particles,
thus allowing the particles to aggregate. At high concentrations of aluminum or
ferric salts, and in the presence of sufficient alkalinity, insoluble hydroxides of
aluminum or iron are formed (see below). In the precipitation reaction, the colloidal particles are enmeshed within the precipitate and thus removed [62–66].

In water treatment, coagulation is defined as a process by which colloidal
particles are destabilized and is achieved mainly by neutralizing their electric
charge. The product used for this neutralizing is called a coagulant.

Flocculation is the massing together of discharged particles as they are brought
into contact with one another by stirring. This leads the formation of flakes or floc.
Certain products, called flocculating agents, may promote the formation of floc.

Separation of the floc from the water can be achieved by filtration alone or by settling [67–71].

Rate of flocculation is dependent upon many factors including concentration of particles, particle contact and range of particle sizes. Coagulation targets dissolved ions such as metal and radionuclides. Some difficulties with this technology include the frequent need to adjust pH levels, the creation of toxic sludge that must be eventually mitigated, and the difficulty that results in trying to address the chemical nature of multiple compounds. This technology has been used consistently in the electronics and electroplating industry as well as for applications in groundwater treatment [63–69].

Coagulation and flocculation are frequently used in the treatment of potable water and preparation of process water used by industry. Certain dissolved substances can also be adsorbed into the floc (organic matter, various pollutants, etc.) [62–64].

2.1.2 Main Coagulants

The most widely used coagulants are based on aluminum or iron salts. The commonly used metal coagulants fall into two general categories: those based on aluminum and those based on iron. The aluminum coagulants include aluminum sulfate, aluminum chloride and sodium aluminate. The iron coagulants include ferric sulfate, ferrous sulfate, ferric chloride and ferric chloride sulfate. Other chemicals used as coagulants include hydrated lime and magnesium carbonate [65–69].

The effectiveness of aluminum and iron coagulants arises principally from their ability to form multi-charged polynuclear complexes with enhanced adsorption characteristics. The nature of the complexes formed may be controlled by the pH of the system.

In certain cases, synthetic products, such as cation polyelectrolytes, can be used. Cation polyelectrolytes are generally used in combination with metal salt, greatly reducing the salt dosage which would have been necessary. Sometimes no salts at all are necessary, and this greatly reduces the volume of sludge produced [65–68].

2.1.3 pH Value for Coagulation and Dosage

Removal of turbidity, suspended solids (SS) and natural organic matter (NOM) using coagulation is well known because of the ability of the process in destabilizing the colloids particles and reducing the repulsion force between the particles. For any water, there is an optimum pH value, where good flocculation occurs in the shortest time with the least amount of chemical. For actual application of coagulating agents, the dosage and optimum pH range should be determined by coagulation control or a jar test [60–63].

2.1.4 Choice of Coagulant

Coagulant should be chosen after the raw water examination in laboratory by means of flocculation test while considering following factors [61–66]:

(a) Nature and quantity of the raw water.
(b) Variations in the quality of the raw water (daily or seasonal especially with regard to temperature).
(c) Quality requirements and use of the treated water.
(d) Nature of the treatment after coagulation (filter coagulation, settling).
(e) Degree of purity of reagents, particularly in the case of potable water.

The chemistry of coagulation/flocculation consists of three processes—flash mix, coagulation and flocculation.

2.2 Sedimentation

When water has little or no movement, SS sink to the bottom under the force of gravity and form sediment. The process by which suspended or coagulated material separates from water by gravity is called sedimentation. Sedimentation is recommended as simple pre-treatment of water prior to application of other purification treatments such as filtration and disinfection methods. It removes undesirable small particulate suspended matters (sand, silt and clay) and some biological contaminants from water under the influence of gravity (Fig. 2.4). In water treatment, it is used to remove solids from waters which are high in sediment content and also to remove particles rendered settle able by coagulation and flocculation [55–58].

The theory of sedimentation would seem to be quite simple. If the settling tank is made large enough and the flow slow enough, this will enhance the rate of fall of the sediment toward the bottom of the tank [69–70].

Sedimentation alone is an effective means of water treatment but is made more effective by coagulation.

Pre-sedimentation basins or sand traps are sometimes used when waters to be treated contain large amounts of heavy SS. This decreases the amount of sediment

Fig. 2.4 A schematic of sedimentation process

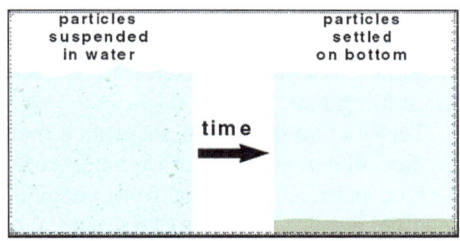

which accumulates in the sedimentation basin as a result of the coagulation and sedimentation process. If water is to be filtered in the course of treatment, coagulation and sedimentation will reduce the load on filters.

2.2.1 Type of Sedimentation Tanks

Sedimentation is used to remove solids from water. It is suitable for water with high sediment content. It is easy to perform and requires a minimum of materials and skill. It can be done with as little as two or more simple storage vessels such as pots and buckets by manual transfer. The effectiveness of a sedimentation tank depends on the settling characteristics of the SS that are to be removed and on the hydraulic characteristics of the settling tank [70–76].

The hydraulic characteristics of a settling tank depend on both the geometry of the tank and the flow through the tank.

Most sedimentation tanks used in water purification today are of the horizontal-flow type. Horizontal-flow tanks may be either rectangular or circular in plan. Circular, horizontal-flow tanks may be either center feed with radial flow, peripheral feed with radial flow or peripheral feed with spiral flow [70–73].

In horizontal-flow tank design, the aim should be to achieve as nearly as possible the ideal condition of equal velocity for all points lying on each vertical line in the settling zone (The ideal basin condition). This in effect would be complete separation of the four zone of the tank.

The sedimentation basins should be equipped with mechanical equipment for continuous removal of settled solids.

2.2.2 Practical Sedimentation Basin

The size and type of particles to be removed have a significant effect on the operation of the sedimentation tank. Because of their density, sand or silt can be removed very easily. The velocity of the water-flow channel can be slowed to less than one foot per second, and most of the gravel and grit will be removed by simple gravitational forces. In contrast, colloidal material, small particles that stay in suspension and make the water seem cloudy, will not settle until the material is coagulated and flocculated by the addition of a chemical, such as an iron salt or aluminum sulfate. The situation in practical sedimentation basins is modified because of the relative density (specific gravity) and shape of the particles, coagulation of particles, concentration of particles and movement of water through the settling tank.

The relative density of suspended matter may vary from 2.65 for sand to 1.03 for flocculated particles or organic matter and mud containing 95 % water.

Floc particles resulting from coagulation with aluminum compounds have a relative density of about 1.18, and those obtained using ferrous sulfate as a coagulant have a relative density of 1.34. These values can be increased by clay or silt or decreased by organic matter. However, most of the particles in a settling basin settle at velocities within Stock's law.

Because of the difference in shape, size and relative density of particles, there is a wide range of settling velocities. This results in some subsiding particles overtaking others, thus increasing the natural tendency of suspended matter to flocculate [68–72].

2.2.2.1 Factors Influencing the Design of Sedimentation Basins

Several factors affect the separation of settle able solids from water. Some of the more common types of factors to consider are

- Sedimentation basins are often designed on the basis of existing installations which are handling the same type of water.
- Experience and judgment of the engineer are also instrumental in the design. However, there are some important points, other than structure, which should be considered in the design of a basin.
- The basin should be large enough to insure an adequate supply of treated water during periods of peak load.
- The characteristics and type of water treatment also affect the design of the basin. Such things as the nature of the suspended material and the amount and type of coagulant needed, if any, should also be considered.
- The influence of temperature is also important, since the viscosity of the water is less on a warm summer day than in cold weather.
- The number of basins depends upon the amount of water and the effect of shutting a basin down. It is desirable to have more than one basin to provide for alternate shutdown of individual basins for cleaning or repairs.
- Basins vary in shape—square, rectangular and round. However, regardless of shape, most basins have slopping bottoms to facilitate the removal of deposited sludge.

The size and type of particles to be removed have a significant effect on the operation of the sedimentation tank.

The shape of the particle also affects its settling characteristics. A round particle, for example, will settle much more readily than a particle that has ragged or irregular edges.

All particles tend to have a slight electrical charge. Particles with the same charge tend to repel each other. This repelling action keeps the particles from congregating into flocs and settling.

Sedimentation basins are equipped with inlets in order to distribute the water uniformly between the basins and uniformly over the cross section of each basin. Inlet and outlets should be designed to avoid short circuiting through the basin [71–77].

2.2.3 Hydraulic Properties of Sedimentation Basin

2.2.3.1 Surface Area or Surface Overflow Rate

The surface area of the tank is one of the most important factors that influences sedimentation. For any particular rate of inflow, the surface area provided determines the tank overflow rate, $v = Q/A$ (see Table 2.1). If there were such a thing as an ideal tank, the tank overflow rate could be made equal to the settling velocity of the particles that the tank was designed to remove.

Because no ideal tank exist, it is customary to reduce the tank overflow rate and to increase the detention time over those indicated by theoretical analysis. It is recommended to apply a correction factor 1–1.25 to both values when settling a discrete solid.

For the sedimentation of flocculent particles from dilute suspensions, the settling velocity will generally be decreased by a factor of 1.25–1.75.

The higher settling velocities or tank overflow rates would be used for warmer water; the lower settling velocities for cold waters.

The settling velocity used in the settling tank design overflow rate is one of the major factors determining tank efficiency [70–78].

2.2.3.2 Depth

The theoretical detention time is equal to the volume of the tank divided by the flow rate. Hence, if A and Q are constant, the theoretical detention time is directly proportional to the tank depth.

As the performance of the tank depends on the flocculation of the SS, and the degree of flocculation depends on the detention time, the tank performance in removal of flocculent particles will depend on its depth.

The efficiency of removal, however, is not linearly related to the detention time. For example, if 80 % of the SS were removed with a detention time of 2 h, a detention time of 3 h might remove only 90 %.

Table 2.1 Typical sedimentation tank overflow rates [5–8]

Type of water	Treatment	Overflow rate ($m^3/m^2 \cdot h$)
Surface water	Alum floc	0.61–0.93
Surface or ground water	Lime softening	0.93–1.54
	Clarification in upflow units	2.44–4.52
		1.83–2.44 (cold water)
		2.44–3.66 (warm water)
	Softening in upflow units	1.83–6.11
		To 3.06 (surface)
		To 4.40 (well)

The raw water entering a sedimentation tank will have a greater density than the water in the tank, as it will contain more SS.

The heavier influent water will tend to form density currents and move toward the bottom of the tank, where it can interfere with the sedimentation process. Density currents are more apt to occur in deep tanks.

Sedimentation basins are commonly designed to remove solids resulting from chemical coagulation of surface water and lime soda as softening of surface and ground waters. In a properly designed basin, a detention time of from 2 to 4 h is usually sufficient to prepare the water for subsequent filtration. When the water is to be used without filtration, longer detention time (up to 12 h) may be provided [73–79].

2.2.3.3 Velocity Through Basin

The velocity of flow through settling basin will not be uniform over the cross section perpendicular to the flow even though the inlets and outlets are designed for uniform distribution. The velocity will not be stable because of density currents and the operation of the sludge removal mechanism. In order to minimize these disturbances, the velocity through a sedimentation tank should be kept between 0.0026 and 0.015 m/s [5–8].

2.2.3.4 Inlet and Outlet Conditions

The inlet to a sedimentation tank should be designed to distribute the water uniformly between basins and uniformly over the full cross section of the tank.

The inlet is more effective than the outlet in controlling density and internal currents, and tank performance is effected more by inlet than by outlet conditions.

The best inlet is one that allows the water to enter the settling tank without the use of pipe lines or channels.

The head loss in preamble baffle ports or basin inlet ports should be relatively large compared to the kinetic energy of the water moving past the permeable ports. This is required to assure equal distribution of flow between tanks and between inlet ports [76–80].

As flocculent solids will frequently be involved, the velocities in the influent channels must be kept low, usually between 0.15 and 0.60 m/s, to prevent break up of the floc. Similar low velocities are required through the inlet ports to reduce the danger of inertial currents interfering with sedimentation.

It has been found that relatively minor changes in an inlet can completely change the hydraulic performance of a settling tank.

The main purpose of the inlet is to provide a smooth transition from the relatively high velocities in the influent pipe to the very low uniform velocity distribution desired in the settling zone, in such a way that interference with the settling process is minimal.

Table 2.2 Typical weir
overflow rates [5–8]

Type of service	Weir overflow rate, $m^3/m\cdot h$
Water clarification	<26
Water treatment	6–7.5
Light alum floc	
(low-turbidity water)	
Heavier alum flow	7.5–11.2
(higher-turbidity water)	
Heavy floc from lime softening	11.2–13.4

The purpose of the outlet is the same except that the transition is from the settling zone to the effluent pipe.

The water level in settling basins is usually controlled at the outlet. This control, however, may be set by means of other than the outlet weir, for example, by a succeeding unit.

It may be desirable to encourage deliberate fluctuation of water level in the settling basins to make use of the storage in them or to break up ice.

Basin outlets are often of the V-notch weir type, and these are quite often provided with means for vertical adjustment to aid in control of overflow. The V-notches help in keeping a uniform flow over the weir at low water levels.

The effect of weir rates, cubic meters per hour per meter of weir, on sedimentation, is not well known, but weir rates are usually limited to commonly accepted values (see Table 2.2).

Circular basins with the inlet at one side and the outlet on the opposite side are not very efficient because of dead areas in the tank and short circuiting of water flow across the tank. The efficiency of circular tanks is much greater if the water is fed to the tank from an inverted siphon located in the center of the tank, and the effluent taken from a weir passing around the entire periphery.

Square basin may be operated in the same manner or may be fed from one side with effluent removed from the opposite side.

The use of baffles in sedimentation tanks should be limited to the inlets and outlets and as remedial measures in poorly designed tanks [77–82].

2.2.3.5 Sludge Handling

The bottom of a settling tank is normally sloped gently toward a sludge hopper where the sludge is collected. The sludge usually moves hydraulically toward the hopper. Sludge scraper mechanisms are used to prevent the sludge from sticking to the bottom and to help its flow.

The sloping bottom and the sludge hopper provide a certain amount of storage space for the sludge before it is removed. The movement of the sludge scrapper mechanism should be quite slow so as not to disrupt the settling process or to resuspend the settled sludge. The velocity of the scrapers should be kept below 18.3 m/h for this reason. Some mechanisms in circular tanks carry vacuum suction pipes instead of squeegees for removing relatively light, uniform solids [78–81].

2.3 Clarification Systems

The equipment used for clarification can be many types; however, the equipment used should provide the correct environment to carry out each step coagulation, flocculation and sedimentation.

Older design for clarification units provided separate chemical addition, flash mixing, flocculating and settling facilities. Modern combined units provide all three steps in one unit, such as sludge recirculation (solids contact) type or sludge blanket unit.

In theory, the sludge blanket unit provides better clarification than the sludge recirculation type as a result of filtering action provided by the sludge bed and the gentle handing of the flocs.

A basic clarification system consists of the clarifier and a chemical feeding system which meters chemical additives in proportion to flow.

The size of standard clarification unit is based upon an upflow rate of approximately 2.5 $m^3/m^2 \cdot h$ (1.2–3.7 range), with a total retention time of 1.5–4 h. The clarified water will contain approximately 5–10 mg/kg of suspended matter [5–8].

Design criteria of clarifiers should be based on steady operation at maximum load. However, it is expected that the actual load will fluctuate over the range from 10 to 100 % of design flow rate and the clarifier should have the capacity to perform satisfactory under these conditions.

Consideration should also be given to anticipation of flow rate limitations and chemical dosages during difficult treatment periods considering high turbidity, low temperature and/or polluted conditions. Manufacturer shall either take such potential difficulties into consideration in design or state the limitations imposed by such conditions [79–83].

2.4 Filters for Water Treating Systems

Multiple units shall be provided to allow continuous operation at full system design capacity with two units out of service (for example, one unit shut down for maintenance and one unit in backwash mode).

Design service flow rates shall be as described in Table 2.3.

Air securing shall be used for units treating effluent at temperatures less than 93 °C. The design air scour rate shall be 90 $m^3/m^2 \cdot h$ minimum. If plant air is unavailable, a separate air compressor shall be included within the system.

Subsurface washers shall be furnished for units treating effluent at temperatures 93 °C and greater. Subsurface wash rate shall be 12.3 $m^3/m^2 \cdot h$, and minimum design backwash rate can be found in the following Table 2.4.

Table 2.3 Flow rates in different types of water filters	Unit	Flow rate (maximum with one unit backwashing) ($m^3/m^2 \cdot h$)
	Downflow, cold pressure type (<65 °C)	9.7
	Downflow, hot pressure type (>65 °C)	10.8
	Downflow, gravity type	9.7

Table 2.4 Minimum backwash rates in different types of water filters	Filter media	Minimum backwash rate $m^3/m^2 \cdot h$
	Sand	36.7
	Anthracite coal	29.2
	Activated carbon	24.5

Bed depth in filters shall be 750 mm minimum. For "in-depth" filters, at least two different density media, of different sizes, shall be furnished.

For hot pressure type units, only washed anthracite coal should be used.

Freeboard shall be a minimum of two-thirds total bed depth, measured from the top of the filter media to the tangent line at the top of vessel.

Filter media traps shall be furnished on the outlet of each pressure filter unit to prevent filter media from entering downstream equipment in the event of under-drain failure. Maximum pressure drop through the trap shall not exceed 35 kPa when the unit is operating at maximum design flow rate. Characteristics of potable water filters should be as per Table 2.5.

Anthracite or marble should be used instead of quartz sand, when any trace of silica must be avoided in industrial process or when they are easier to obtain.

2.5 Quantities of SS which can be Removed by Filtration

The following consideration should be made as a guiding principle:

The SS lodge between the grains of the filter material. Since sufficient space should always be left for the water to percolate, the sludge should not, on average, fill more than one quarter of the total volume of voids in the material.

Irrespective of grain size, one cubic meter of filtering material contains about 0.45 m^3 of voids, the volume available for the retention of particles is about 0.11 m^3, provided that the effective grain size of the filtering medium is suitable to the nature of the particles [80–84].

When the SS are based on colloidal floc, their dry matter content does not exceed 10 kg/m^3; the quantity that can be removed per m^3 of filter material is therefore no more than $0.11 \times 10 = 1.1$ kg.

Table 2.5 Characteristics of potable water filters [5–8]

Filter type	Permissible filtration rates (m³/m²·h)		Design pretreatment to reduce turbidity in applied water to (mg/kg)		Head required (m)		Length of filter run (h)		Min. thickness (mm)	
	Maximum day	Maximum rate	Average	Maximum	Clean filter	Maximum	Average	Minimum	Gravel	Sand
Rapid sand gravity	4.9	12.2	2	5	0.3	2.4	36	5	304.8	508
Pressure	4.9	12.2	2	5	0.3	7.6	48	5	304.8	609.6
Slow sand	2.4	7.3	1	3	0.6	1.2	1,000	250	304.8	1,066.8
Diatomite	2.4	7.3	1	3	2.1	21.3	6	0.5	304.8	

2.6 Process Used for Boiler Feedwater Treatment

Vessel should be of sludge blanket type employing a central downcomer. Separate (not integral) deaerator compartment is preferred. The treater should be so sized that the rising rate of settled water is such that effluent treated water has a turbidity of less than 10 mg/kg.

Separate clean and dirty backwash compartments shall be sized to meet normal filter backwash and sodium zeolite regeneration requirements without increasing flow rate through the unit to more than 10 % of normal design.

The clean backwash water compartment should be replenished by filtered water at a much slower rate than the backwash rate.

The dirty backwash water should be returned at a set rate, so that heat and water are recovered. The treatment should be carried out at low pressure corresponding to vapor pressure for temperatures chosen between 102 and 115 °C as required. Units shall be designed for continuous service and uninterrupted operation for a period of 2 years.

All equipment shall be suitable for unsheltered outdoor installation for the climatic zone specified. The total detention time of the vessel should not be less than 90 min at rated capacity of flow [81–85].

The maximum allowable upflow rate (rinse rate) through the unit shall be 3.7 $m^3/m^2 \cdot h$ at water temperatures above 90 °C. This rate shall not be exceeded when backwashing filters rinsing softeners.

This rate shall be reduced to 1.0 $m^3/m^2 \cdot h$ or less for waters containing appreciable organic matter turbidity or magnesium to meet guaranteed effluent turbidity of less than 10 nephelometric* units for a range of 10–100 % of design raw water throughput. Chemical mix tanks and pumps should be provided for hot-process treater. Incoming water should be provided at pressure sufficient to overcome the following losses [5–8]:

- pipe friction;
- static head to the top of the softener;
- vent condenser;
- spray nozzle;
- water flow meter;
- water level control valve; and
- vessel operating pressure (exhaust steam pressure).

2.7 Ion Exchange

Ion-exchange resins are used to replace the magnesium and calcium ions found in hard water with sodium ions. When the resin is fresh, it contains sodium ions at its active sites. When in contact with a solution containing magnesium and calcium ions (but a low concentration of sodium ions), the magnesium and calcium ions preferentially migrate out of solution to the active sites on the resin, being replaced in solution by sodium ions. This process reaches equilibrium with a much lower concentration of magnesium and calcium ions in solution than was started with. Ion-exchange resins are used to remove poisonous (e.g., copper) and heavy metal (e.g., lead or cadmium) ions from solution, replacing them with more innocuous ions, such as sodium and potassium.

2.7.1 Classification of Ion Exchange Resins

Ion exchange resins are classified according to their specific application as per Table 2.6.

2.7.2 Design Criteria for an Ion Exchange System

Design criteria for an ion exchange system should be based upon [5–8]:

- the required flow rate;
- influent water quality,

Table 2.6 Classification of ion exchange resins [5–8]

Type	Application	Ionic form in the ready-to-use condition	Regenerating agent. Aqueous solution of
Cation exchange resins strongly acidic	Reduction of calcium ion concentration	Na	NaCl
Cation exchange resins strongly acidic	Reduction of salt content	H	HCl, H_2SO_4
Weakly acidic	Reduction of hydrogen carbonate concentration	H	HCl, H_2SO_4, CO_2
Weakly acidic	Reduction of heavy metal ion content	Na, H	HCl, H_2SO_4, NaOH
Anion exchange resins strongly basic	Reduction of salt content	OH	NaOH
Anion exchange resins strongly basic	Reduction of the content of certain ions, for example, nitrate ions, sulfate ions	Cl, HCO_3	NaCl, $NaHCO_3$
Anion exchange resins strongly basic	Reduction of the organic substance content, for example, humic acids	Cl, OH	NaCl, NaOH
Weakly basic	Reduction of salt content	Free base	NaOH
Weakly basic	Reduction of heavy metal ion content	Free base	NaOH
Weakly basic	Reduction of the organic substance content, for example, humic acids	Free base	NaOH

- desired effluent water quality;
- exchange capacity and hydraulic characteristics of the exchanger;
- period between regenerations;
- type of operation manual or automatic; and
- flexibility required, that is, the number of softener units.

The ion exchangers are not economically suitable for demineralizing waters containing more than 1,000–2,000 mg/kg of dissolved solids, except in a few specialized industrial applications.

The process of ion exchange for softening waters is preferable to precipitation process when one or more of the following conditions exist:

- less than 100 mg/kg of hardness expressed as calcium carbonate is present in the water;
- an extremely low dissolved solids content is required;
- only a limited volume of treated water is required.

Relative exchange capacity of cation exchangers and regenerative salt dosage would be as per Table 2.7.

The anion exchangers have typically an exchange capacity calculated as $CaCO_3$ of 27.4–57.2 g/L at a sodium dosage of 1.05–7 kg/kg removed. Interstate degasification in demineralization systems should be considered at flow over 22.7 m^3/h and alkalinity over 100 mg/kg.

When ultra-pure water is required, using the mixed bed demineralizer is recommended.

The demineralized water storage(s) shall be designed in order to store the produced demineralized water and to cover the following users:

- Make-up to deaerators.
- Process units.
- Regeneration of condensate treatment.

Table 2.7 Relative exchange capacity of cation exchangers [5–8]

Cation exchanger	Nominal exchange capacity (g/L)	Regenerative salt dosage	
		Volumetric (kg/m^3)	Effective (kg/kg) hardness removed
Greensand	6.4	20.2	3.1
Processed greensand	12.6	39.5	3.1
Synthetic siliceous zeolite	25.2	79.2	3.1
Resin, polystyrene	73.2	201.7	3.1
Resin, polystyrene	50.3	80	1.7

2.8 Standard Specification of Demineralizing Unit

This section specification covers the general requirements for the design, construction and inspection of automatic regenerating type demineralizing units for the production of boiler feedwater.

2.8.1 The Demineralizing Unit Design

The demineralizing unit shall consist of but not necessarily be limited to the following equipment:

- Cation resin bed for exchanging acidic hydrogen.
- Anion resin bed for exchanging basic hydroxide.
- Degasifier (decarbonator) removing carbon dioxide formed in the cation resin bed, if required.
- Mixed bed polisher containing strong cation and anion resin for exchanging acidic hydrogen and basic hydroxide, respectively, if required.
- Regenerating equipment including chemical storage tanks, measuring tanks, pumps, blowers, instrumentation for control, neutralizing equipment for regeneration-effluent, interconnecting piping and others [76–82].

2.8.2 Chemical for Resin Regeneration

The following chemicals shall be used for the regeneration of resin:

- H_2SO_4 or HCl solution for cation exchanger.
- NaOH solution of anion exchanger.
- Skid-mounted chemical storage tanks shall be provided and equipped with a chemical transfer pump for regeneration purposes.
- The capacity of each chemical storage tank shall be designed to enable maximum operation within five cycles.

Two separate chemical transfer pumps shall be provided, for each chemical. These chemical pumps shall be driven by individual motors, one for sulfuric acid transfer and the other for caustic solution transfer.

The unit shall be designed to minimize consumption, and its guaranteed values shall be satisfied [63–75].

2.8.3 Demineralized Water Quality

The following data on demineralized water quality shall be specified [5–8]:

- Electrical conductivity.
- Total hardness.
- Silica.
- pH.
- Other requirements.

2.8.4 Type of Demineralizing Unit

Type of demineralizing unit is to be decided, based on the raw water analysis. The use of a two-bed two-tower, two-bed three-tower unit or others suitable for the specified raw water quality and treated water quality shall be considered.

- The unit shall continuously produce a net flow to the service. However, where adequate demineralized water storage is available to meet standby and service requirements, a single unit may be permitted.
- Expected turndown ratio of a demineralized water flow rate to a designed value shall be suitable for boiler feed operation, as specified.

The demineralizing unit shall be installed in a non-hazardous area. Therefore, the unit shall be designed for suitable outdoor installation.

Cation/anion resin vessels shall be made of carbon steel with an inner rubber lining or equivalent to protect against corrosion.

Where diluted sulfuric acid and caustic soda are used, vessel internals shall be made of rubber-lined carbon steel, Type 316 stainless steel or equivalent [75–84].

2.8.5 Performance Characteristics

- The following performance characteristics shall be guaranteed:
- Treated water output capacity per hour and per cycle.
- Inlet water flow rate of both operation and regeneration.
- Treated water quality per items:
- Electrical conductivity.
- Silica.
- Total hardness.
- pH at 250 °C.
- Operating/regeneration cycle time.
- Chemicals both for regeneration and for neutralization per cycle (kg/cycle) and per each treated water (kg/m^3).
- Waste water quantity per cycle (m^3/cycle) and per each treated water (m^3/m^3) [68–73].

2.9 Miscellaneous Processes

Reverse osmosis (RO) is a membrane-technology filtration method that removes many types of large molecules and ions from solutions by applying pressure to the solution when it is on one side of a selective membrane. The result is that the solute is retained on the pressurized side of the membrane and the pure solvent is allowed to pass to the other side.

2.9.1 Reverse Osmosis

The applied pressure for brackish water purification is typically in the range of 2,760–4,140 kPa (ga) [27.6–41.4 bar (ga)] and for seawater purification, in the range of 5,520–6,900 kPa (ga) [55.2–69.0 bar (ga)].

Recovery of product (desalted) water with reverse osmosis units ranges from 50 to 90 % of the feedwater depending upon the feedwater composition, the product water quality requirement and the number of stages utilized.

For water containing from about 250–1,500 mg/kg dissolved solids, an economic comparison of ion exchange and reverse osmosis is recommended to select the more cost effective process.

Reverse osmosis may be considered for desalination of seawater.

In many cases, the reverse osmosis product water shall be treated by one of the ion exchanger processes, if high quality feedwater is required.

A pre-treatment system shall be provided to avoid fouling or excessive degradation of the membrane. Typically pre-treatment will include filtration to remove suspended particles and addition of chemicals to prevent scaling and biological growth.

Heating feedwater to provide optimum operating temperature of 25 °C for reverse osmosis system shall be considered.

Process design of reverse osmosis system shall be based on feedwater and product water qualities and rates. Different types of reverse osmosis modules layouts, for example, parallel, series including reject staging and product staging shall be proposed by Vendor(s) and the final configuration will be selected upon Company's approval [62–71].

2.9.2 Electrodialysis

Recovery of product (deionized) water with electrodialysis units ranges from 50 to 90 % of the feedwater depending upon the number of stages and degree of recirculation utilized.

Operating cost consists mainly of power costs (typically 1.6–2.7 kWh/m^3 of product water) and membrane cleaning and replacement costs.

Based upon combined capital and operating costs, the electrodialysis process is most economical when used to desalt brackish water (1,000–5,000 mg/kg dissolved solids) to a product water concentration of about 500 mg/kg dissolved solids. Process design of electrodialysis unit shall be based on feedwater and product water qualities and rates.

Chapter 3
Raw Water Systems

Keywords Design of water system · Disinfection · Superchlorination · Dechlorination · Ozonation · Activated carbon · Raw water · Plant water

Basis for the water supply in oil, gas and chemical industries is the availability of raw water from wells or surface water and their subsequent treatment as cooling water and/or as supplementary water for the cooling water and water/steam circuit. During freshwater cooling, the water must be freed from pollution which occurs in the form of floating or suspended materials. No matter where your makeup water comes from, it may contain contaminants that can foul or damage downstream equipment and affect the final process water quality [86–89].

Surface water, well water or reclaimed water can contain suspended solids, colloidal matter, organics, hardness, silica, iron, manganese and other contaminants. That is why it is so important to choose the proper water pre-treatment system for your process.

Groundwater may be the preferred source, but the most convenient source of water for plants is frequently a natural stream or river close by. The two most important criteria in judging the suitability of the surface water source are the quality of the water and the reliability of the flow.

Surface water accumulates mainly as a result of direct runoff from precipitation (rain or snow). Precipitation that does not enter the ground through infiltration or is not returned to the atmosphere by evaporation flows over the ground surface and is classified as direct runoff. Direct runoff is water that drains from saturated or impermeable surfaces, into stream.

In tropical countries, rivers and streams often have a wide seasonal fluctuation in flow. This also affects the quality of the water. In wet periods, the water may be low in dissolved solids concentration but often of a high turbidity. In dry periods, river flows are low and the load of dissolved solids is more concentrated.

Mountain streams sometimes carry a high silt load, but the mineral content is mostly low and human pollution is generally absent. In plains and estuaries, rivers

A. Bahadori et al., *Essentials of Water Systems Design in the Oil, Gas, and Chemical Processing Industries*, SpringerBriefs in Applied Sciences and Technology, DOI: 10.1007/978-1-4614-6516-4_3, © The Author(s) 2013

usually flow slowly except when there is a flood. The water may be relatively clear, but it is almost always polluted, and extensive treatment is necessary to render it fit for drinking and domestic purposes. The quality of river water does not usually differ much across the width and depth of the riverbed [87–90].

3.1 Design of Water Systems

Raw water intakes withdraw water from a river, lake or reservoir over a predetermined range of pool levels. Screens remove large floating objects from the water to protect pumping equipment. Aeration removes gases and volatile compounds and also oxidizes certain dissolved metals. Intake designs aim to avoid clogging and scouring and to ensure the stability of the structure even under flood conditions.

The entrance of large objects into the intake pipe should be prevented by the use of a coarse screen or by obstructions offered by small opening in the cribwork or riprap placed around the intake pipe.

The area of the openings in the intake crib should be sufficient to prevent an entrance velocity greater than about 0.15 m/s, in order to avoid carrying settleable matter into the intake pipe. Intake ports may be placed at various elevations so that water of the best quality may be taken. They should also be placed so that if one or more of the ports are blocked, another can be opened.

Fine screens for the exclusion of small fish and other small objects should be placed at an accessible point, as at the suction or wet well at the pumping station where the screens can be easily inspected and cleaned.

Submerged ports should be designed and controlled to prevent air from entering the suction pipe. The difficulty can be minimized by maintaining an entrance velocity not greater than 0.15 m/s, preferably much less than this, and maintaining a depth of water over the port of at least three diameters of the port opening. The capacity of the intake should be sufficient for future demands during the life of the structure.

Parallel bars, preferably removable, may be placed over intake ports with openings between bars not less than 25–50 mm. The grids have sometimes been electrically charged to keep fish away [88–91].

Self-cleaning screens of moving-belt type over intake ports are in successful use. Vigorous reversal of flow through the intake port and screen or grid is sometimes created as an expedient for cleaning. Provision should be made for such reversal in the design of the intake structure and conduit.

When selecting pumps, the design consultant considers the following items:

- Select a pump-operating curve where the required operating point is beyond the minimum and near the maximum efficiency point (optimally just to the right of this point) of the pump curve.
- Select a pump-operating curve where the operating point is near the minimum value of radial thrust.

- Specify a pump/impeller located near the center of the pump-operating curve recommended operating range, to facilitate modifying the pump with a different impeller to change pumping performance. This modification may be required based on information determined during station start-up operational testing when the pump is discharging into the system.
- In specifying the pump-operating point, specify an operating flow at the required head, that is, 105 % of the design requirement to allow for losses.
- Operating capacity from pump wear and increased pipe friction (design flow = peak design × 1.05).

The conduit conveying water from the intake should lead to a suction well in or near the pumping station. Either a pipe, lying in or buried in the bottom of the body of water, or deep tunnels may be used as intake conduits.

The capacity of the conduit and the depth of the suction well should be such that the intake ports to the pumps will not draw air.

A velocity of 0.6–0.9 m/s in the intake conduit, with a lower velocity through the ports, will give satisfactory performance.

The horizontal cross-sectional area of the suction well should be three to five times the vertical cross-sectional area of the intake conduit. Pumps should be started gradually to avoid drawdown in the suction well, and they should be stopped gradually to prevent surge.

The intake well acts as a surge tank on the intake conduit, thus minimizing surge.

The intake conduit should be laid on a continuously rising or falling grade to avoid accumulation of air or gas, pockets of which would otherwise restrict the capacity of the conduit.

Where air traps are unavoidable, provision should be made to allow gas to be drawn off from them. Where pipes are used, they should be weighted down to avoid flotation.

It is recommended that the intake works be in duplicate because of the almost complete dependence of the waterworks on its intake, the intake conduit and the suction pit. Two or more widely separated intakes are highly desirable.

An aqueduct is a channel or pipe used to transport water from a remote source to a desired location, such as a town, city or agricultural area. It is designed to convey water from a source to a point, usually a reservoir, where distribution begins. An aqueduct may include canals, flumes, pipe lines, siphons, tunnels or other channels, either open or covered, flowing at atmospheric pressure or otherwise.

The choice between available types of conduits in an aqueduct depends on topography, available head, quality of water and possibly other conditions.

Water in aqueducts should be protected against pollution by infiltration of non-potable groundwater, the overflow into the aqueduct of polluted surface waters, and all other possible sources of pollution to which water flowing low or atmospheric pressure may be exposed.

Part of the precipitation that falls infiltrates the soil. This water replenishes the soil moisture or is used by growing plants and returned to the atmosphere by transpiration. Water that drains downward (percolates) below the root zone finally reaches a level at which all the openings or voids in the earth's materials are filled with water. This zone is called the zone of saturation. The water in the zone of saturation is called the groundwater. Ground waters are, generally, characterized by higher concentrations of dissolved solids, lower levels of color, higher hardness (as compared to surface water), dissolved gasses and freedom from microbial contamination.

It is important to keep water entrance velocity through the screen openings between 0.03 and 0.06 m/s. Such velocities will minimize head losses and chemical precipitation or incrustation. Care should be taken in estimating the effective screen area. It is not uncommon to allow for as much as 50 % plugging of the screen slots by formation particles.

The total open area required should be obtained by adjusting either the length or diameter of the screen because the slot is not arbitrary. It is sometimes advisable to construct an artificial gravel-pack well to permit an increase in screen slot size. When a well screen is surrounded by an artificial gravel wall, the size of the openings is controlled by the size of gravel used and by the type of openings.

Actual screen design should not be final until samples of the aquifer front, the actual well location, are available for proper sieve analysis. However, experience and samples from test wells in the area permit preliminary calculations of the openings. The actual open area per meters of screen depends upon the type of construction and the manufacturer [91–95].

Pumping stations for obtaining water from a water source have two types depending on the source:

- pumping from surface water (river, canal, lake, reservoir, etc.);
- pumping from subsurface water (soil water, deep-seated spring, cavern water, spring water, marginal water, etc.).

Conditions to be considered in the location of waterworks pumping station include the following:

- Sanitary protection of the quality of water;
- The hydraulics of the distribution system;
- Possibilities of interruption by fire, flood or other disaster;
- Availability of power or of fuel;
- Growth and future expansion.

Floods offer a hazard which may be minimized by favorable location and site protection. To obtain the greatest hydraulic advantage, the station should be located near the middle of the distribution system.

The danger of interruption of service by fire should be considered, and the station should be protected by fire walls, fireproofing and a sprinkler system. Pumping stations may sometimes be located underground when conditions are favorable.

Underground locations require care to avoid flooding and dampness. Attention should be given to accessibility illumination, ventilation and heating and providing adequate space for operation and maintenance [90–94].

The type of primary or of auxiliary power selected should be the most reliable, the most available and the least expensive. If all three conditions cannot be fulfilled, they should be rated in the order stated, with the greatest reliability as the most important.

Auxiliary sources of power include internal combustion engines, gas turbines, steam engines or a secondary source of electric energy.

There should be sufficient power in a waterworks pumping station to supply the peak demand without dangerously overloading the power equipment. The highest efficiency and the greatest economy of operation require that the total load on all the units in the pumping station should be divided among them in such a manner that each can operate at its rated capacity.

Standby pumping equipment should be provided in pumping stations. One or more horizontal shaft centrifugal pumps may be equipped with an electric motor on the pump axis on one side of the pump and an internal combustion engine on the other side, or two electric motors, driven from different circuits, may be placed on the same shaft.

The use of compressed air in pumping stations should be considered as an auxiliary power for starting engines, blowing boiler tubes, operating control systems, pumping wells and other purposes.

The layout of the piping in a pumping station may be as important as the efficiency of operation in affecting economy. Short, straight, well-supported pipe lines, devoid of traps for sediment or for vapor, sloping in one direction to drains and with adequate cleanouts, should be the object of the designer.

A few details to be observed in making connections to centrifugal pumps are as follows:

- the elbows on the suction side of a double-suction pump should be normal to the suction nozzle, or a special elbow with guide vane followed by a piece of straight pipe to the suction nozzle should be used, to equalize the flow on each side of the impeller;
- on the pump discharge, where change in direction is necessary, a constant-diameter bend should be used, followed by a long, straight reducer or increaser. Such an installation will most effectively interchange velocity and pressure heads;
- eccentric reducers with the top of the pipe and the top of the reducer at the same level should be used in horizontal suction piping to avoid the creation of an air pocket;
- an increase whose length is ten times the difference in diameters may be considered "long".

Pumping stations can be grouped as follows:

- pumping water from a water source such as a river;
- for lifting water (high quantity, low pressure) from a well;
- for pumping water into a supply system, elevated water tank or water tower to increase pressure.

Raw water pumped from sea intake after treatment in desalination plant would be stored as desalinated water in storage tanks.

Raw water pumped from outside the refinery after treatment for surface water or usually without any treatment for groundwater would be stored as clarified water in storage tanks.

Clarified (or desalinated) water stored in the clarified (or desalinated) water tanks is pumped to the plant water header for refinery and/or plant use, makeup water for the cooling tower, makeup water for the boiler feedwater treatment plant, feed to the potable water sand filters and makeup for fire water tanks.

Sea water after desalination shall be considered as makeup for fire water, machinery cooling water and demineralization plant.

Sea water shall be considered in emergency as alternate source for fire water.

After branching from main header of plant water for fire water tanks and potable water system, one backflow preventer should be installed on the header to insure against backflow of possibly contaminated water from process units into lines used for drinking and sanitary services.

Potable water is clarified water that has been filtered in sand filters and chlorinated. Two potable water systems should be provided, one for refinery and or plant use and the other for the employee housing with identical flow scheme, if required.

The scheme of each system with different capacities should consist of sand filtration, chlorination and storage tanks. A cross-connection should be provided to furnish water in an emergency from the discharge of refinery and/or plant potable water pumps to the housing potable water tanks. Each system should have a separate chlorination system. Design of chlorination system is presented briefly below [92–97].

3.2 Disinfection

Disinfection is usually the final stage in the water treatment process in order to limit the effects of organic material, suspended solids and other contaminants.

The purpose of disinfection is to render water safe for human consumption, free from pathogenic bacteria and therefore incapable of transmitting disease. Water disinfection means the removal, deactivation or killing of pathogenic microorganisms. Microorganisms are destroyed or deactivated, resulting in the termination of growth and reproduction.

Chlorine, in its various forms (liquid, gas or hypochlorite), is the major chemical widely used in disinfecting water. Chlorine is widely used for disinfection of drinking water and also for general disinfection and decontamination. It is cheap, widely available and effective. Other disinfectants are iodine, bromine, ozone, chlorine dioxide, ultraviolet light and lime, which might be considered if chlorine gas is not readily available. Figure 3.1 is a typical diagram of raw water treatment facilities.

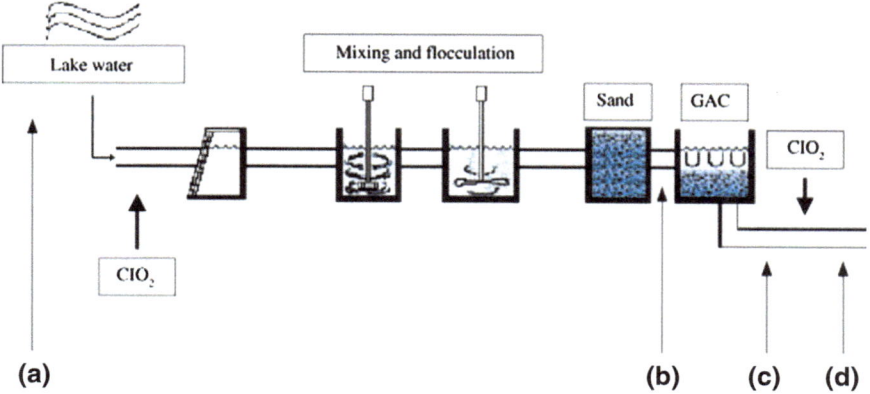

Fig. 3.1 Schematic diagram of a raw water treatment plant. **a** Raw water. **b** Water after pre-disinfection with chlorine dioxide, flocculation and sand filtration. **c** Granular-activated carbon-filtered water. **d** Tap water after post-disinfection with chlorine dioxide [88]. © Elsevier, 2005, reprinted with permission

Chlorination chemicals are relatively:

- Easy to obtain
- Economical
- Effective
- Easy to apply

Typical forms of chlorine used in wastewater treatment are as follows:

- Elemental chlorine
- Hypochlorite
- Chlorine dioxide

The important applied disinfectants are as stated below:

While the principal use of chlorine is as a disinfectant, its application is practiced for prevention and destruction of odors, iron and/or color removal simultaneously. Chlorination is classified according to its point of application and its end result.

Whenever surface waters are used with no other treatment than chlorination, its role is extremely important as the principal if not the only safeguard against disease.

Such otherwise untreated waters are likely to be rather high in organic matter and require high dosages and long contact periods for maximum safety. The chlorine can be added to the water in the pipe leading from an impounding reservoir to the residential township. For disinfection alone, a dose of 0.5 mg/L or more may be required to obtain a combined available residual in the township distribution system.

Apart from diatomaceous earth filters for which pre-chlorination is a must, in case of all kinds of rapid filters, the chlorine may be added in the suction pipes of raw water pumps or to the water as it enters the mixing chamber. Its use in this manner may improve coagulation and reduce tastes and odors, may keep the filter sand cleaner and may increase the length of filter runs. Frequently, the dosage is such that a combined available residual of 0.1–0.5 mg/L goes to the filters.

This usually refers to the addition of chlorine to the water after all other treatments. It is standard treatment at rapid sand filter plants, and when used without pre-chlorination and with low residuals, it is sometimes called marginal chlorination. The chlorine may be added in the suction line of the service pump, but it is preferable to add it in the filter effluent pipe or in the clear well so that an adequate contact time will be assured. This should be at least 30 min before any of the water is consumed if only post-chlorination is given.

Chlorine dosage may be established from either bench-scale laboratory testing, or actual measurement of field results from known plant operation. The results are suitable for establishing base feed rates; however, real-time corrections must be made to adjust for changing conditions. Since field conditions are not as controlled as laboratory tests, the actual dosage will generally be higher than those established in the laboratory.

Dosage will depend upon the character of the water and may be 0.25–0.5 mg/L in order to obtain a combined available residual of 0.1–0.2 mg/L as the water leaves the plant.

Greater residuals will probably be needed if it is desired to hold a disinfecting effect throughout the distribution system [95–99].

Breakpoint chlorination is related to the chlorine necessary to satisfy the inorganic, ammonia and organic demands of the wastewater. Once achieved, additional chlorine applied to the wastewater is in the form of free chlorine.

The breakpoint indicates complete oxidation of the available ammonia and any other organic amines, and the residual above the breakpoint is mostly free available chlorine. Usually, the chlorinous and other odors will disappear at or before the breakpoint. Dosage is likely to be 7–10 mg/L in order to obtain a free available residual of about 0.5 mg/L or more. The chlorine, when the breakpoint procedure is applied, usually, but not always, is added at the influent to the plant. In some cases, ammonia has been added to water lacking in it in order to form a more pronounced breakpoint.

The chlorine combines with the ammonia and other organic amines to form chloramines. Chloramines are disinfectants used to treat drinking water. Chloramines are most commonly formed when ammonia is added to chlorine to treat drinking water. The typical purpose of chloramines is to provide longer-lasting water treatment as the water moves through pipes to consumers. This type of disinfection is known as secondary disinfection. They are less active than hypochlorous acid (HOCl), and their disinfecting efficiency is considerably less than HOCl, but its bactericidal effects are maintained over a longer period. The beneficial aspect of ammonia chlorine treatment is that little or no combination of chlorine with organic matter occurs, to produce undesirable odors. Hence, chloramines have been used as an odor preventive with satisfactory bactericidal effects, if longer contact periods are provided. Ammonia is usually added in the ratio of 1 part of ammonia to 4 parts of chlorine, although experiment may indicate the desirability of a higher or lower proportion. It is used as the gas, as a solution of the gas in water or as ammonium sulfate or ammonium chloride. The ammonium sulfate used is the chemical fertilizer, which is sufficiently pure to be used for this purpose.

If taste and odor control is the main object of chloramination, the ammonia should be added in advance of the chlorine [99–102].

3.3 Superchlorination and Dechlorination

Superchlorination–dechlorination is a chemical water treatment method that requires two steps. This highly effective chemical water treatment requires an initial dose of chlorine in high concentrations. In the second step, peroxide is added to remove the chlorine. This chemical water treatment method is considered inconvenient, but highly effective in disinfecting water.

The process of superchlorination followed by dechlorination is best defined as the application of chlorine to water to produce free residual chlorination, in which the free available chlorine residual is so large that dechlorination is required before the water is used. Whenever highly polluted waters have to be disinfected for drinking purposes, adoption of this process is recommended.

Where superchlorination or alternatively breakpoint chlorination is practiced, it is desirable to provide a baffled contact tank of approximately 30 min retention and in no case less than 20 min, in order to ensure complete sterility. The most common dechlorinating agent is sulfur dioxide, which is added either at the end or in the last day of the contact tank so as to ensure complete mixture and dechlorination to leave the desired residual.

It is normally found that the dose of sulfur dioxide is approximately 20 % greater than the theoretical amount calculated to combine with the chlorine removed. In certain small supplies, sodium thiosulphate or sodium bisulfite has been employed as dechlorinating agents. If complete dechlorination is desired, then passage of the highly chlorinated water through a bed of granular carbon is often used. Very occasionally, a small dose, 0.2–0.5 mg/kg (mass ppm or ppm by mass), of potassium permanganate is also added to the contact tank prior to filtration in order to control tastes [97–101].

For injection of chlorine gas into water, a variety of chlorinators have been designed and manufactured. Chlorine gas is obtained in pressurized cylinders ranging from 45 to 1,000 kg capacity. Small plants commonly use the 45 kg cylinders, while large plants requiring 75–100 kg/day generally use 1,000 kg containers as a matter of convenience and economy.

The satisfactory chlorinators must feed the gas into the water at an adjustable rate, and it must do this although the pressure in the gas container changes as the temperature changes. While some chlorinators apply the measured amount of gas to the water through a porous porcelain diffuser, most types dissolve the gas in water and feed the solution.

Certain precautions are needed in chlorination practice. Chlorinators and cylinders of chlorine should be housed separately and not in rooms used for other purposes. Ventilation should be provided to give a complete air change each minute, and the air outlet should be near the floor since chlorine is heavier than air. Switches for fans and lights should be outside the room and near the entrance. The entrance door should have a clear glass window to allow observation from outside. Chlorinator rooms should be heated to 15 °C with protection against excessive heat. Cylinders should be protected against temperatures greater than that of the equipment.

Chlorinated lime has been largely displaced, not only by chlorine gas, but also by improved commercial compounds of sodium and calcium hypochlorite.

Hypochlorination is especially applicable to emergency use where supplies are endangered, and there would be considerable delay in obtaining chlorine gas and chlorinators [92–98].

Chlorination has been proved to be responsible for increases in the concentration of volatile halogenated organics, particularly chloroform, bromodichloromethane, dibromochloromethane and bromoform, which are commonly found in chlorinated water.

3.4 Ozonation

Ozone is a strong oxidizing agent and may be applied in any situation where chlorine has been used. Ozone has greater disinfection effectiveness against bacteria and viruses compared with chlorination. In addition, the oxidizing properties can also reduce the concentration of iron, manganese and sulfur and reduce or eliminate taste and odor problems. Ozone oxides the iron, manganese, and sulfur in the water to form insoluble metal oxides or elemental sulfur. These insoluble particles are then removed by post-filtration. Organic particles and chemicals will be eliminated through either coagulation or chemical oxidation. Ozone is unstable, and it will degrade over a time frame ranging from a few seconds to 30 min. The rate of degradation is a function of water chemistry, pH and water temperature.

Dosages range from 0.25 mg/L for high-quality groundwater to 5 mg/L following filtration for poor-quality surface waters.

Effective ozone dosages for viruses range from 0.25 to 1.5 mg/L at contact times of 45 s to 2 min. Ozone, unlike chlorine and the other halogens, is not particularly sensitive to pH within the range of pH 5 to 8, but is significantly affected by temperature.

The disadvantages of ozonation, which have restricted its use worldwide, are its cost relative to chlorine, the need to generate it at the point of use and its spontaneous decay which prevents maintenance of a residual in the distribution system. However, the fact of production of halogenated hydrocarbons by present popular chlorination practice would make the said disadvantages of lesser importance, especially if production of ozone should be or is generated for other needs [91–98].

3.5 Activated Carbon

Activated carbon is used in water processing, primarily as a short-term treatment to correct seasonal taste and odor problems.

Powdered activated carbon is generally less than 0.075 mm in size and thus has an extremely high ratio of area to volume.

It is applied as slurry at points of raw water entry, at the mixing basin, split feed, with a portion in the mixing basin and the balance just ahead of the filters, either at constant rate or at a heavy rate immediately after the filter is washed, followed by a light rate (during seasonal time period).

The dosages used vary from 0.25 to 8 mg/L with 7–2 mg/L most common.

The carbon can be used in impounding reservoirs also to reduce algae or either odors. In this case, it should be applied as slurry and sprayed over the water surface at 1–10 g/m^2.

The required dosage of activated carbon can be controlled by means of the threshold odor test [93–101].

Chapter 4
Water Pollution

Keywords Water pollution · Pollutant sources · Effluent water · Pollution control · Spill prevention · Groundwater pollution · Biological treatment · Ecosystems · Fertilizer

In areas where oil and gas development is prevalent, air, water and soil resources can become contaminated with oil and gas wastes and by-products. The petroleum industry has been concerned with water use and the subsequent handling and treatment of wastewater for many years [103–106].

Prevention of spills of oil and related petroleum products should be one of the prime objectives, both in the design and the operation of the proposed oil, gas and chemical facilities, and should include, but not be limited to siting and design criteria for all facilities, operating procedures and their periodic review, inspection and monitoring of facilities, personnel training, revision of operating procedures (where required), and redesign of facilities (if necessary). This chapter is intended to cover the safety and environmental control aspects as the minimum requirements for water pollution control in oil, gas and chemical processing and production plants.

Each year, an average of about 5 million tons of petroleum is transported across the seas around the world [105] putting the marine lives and ecosystem in a dire risk.

Hence, the impact of oil spill on the ecosystem is severe and cannot be overemphasized (Fig. 4.1). Spills affect marine life. Marine birds, especially diving birds, and shell fishes are the most vulnerable (Fig. 4.1a). However, the effect of chemical dispersants most commonly used to control the spills may even be more harmful and in some cases kill shell fishes. Oil spills also spoil beaches (Fig. 4.1d) and shorelines (Fig. 4.1c) [106].

A. Bahadori et al., *Essentials of Water Systems Design in the Oil, Gas, and Chemical Processing Industries*, SpringerBriefs in Applied Sciences and Technology, DOI: 10.1007/978-1-4614-6516-4_4, © The Author(s) 2013

Fig. 4.1 a An oil-stained pelican where hundreds of pelican nests exist (AP Photo/Gerald Herbert). **b** Fireproof boom used to contain in situ burning (Office of Response and Restoration, National Ocean and Atmospheric Administration). **c** A shrimp boat (AP Photo/Eric Gay) and (John Moore/Getty Images). **d** A beach soiled with oil [107], © Elsevier, 2010 reprinted with permission

4.1 Water Pollution Terminals

4.1.1 Wastewater Pollutant Sources Crude Oil Terminal

The onshore facilities for most crude terminals will consist of storage tanks and associated equipment for crude oil, ballast water and sanitary water. Thus, the major environmental concern is contamination of wastewaters with oil and the treatment of the ballast and sanitary waters prior to discharge. The treating methods for oil-contaminated wastewaters include various types of separators. It is most beneficial to segregate the dirty and clean waters and thereby to minimize the volume of water requiring treatment [104–108].

4.1.2 Product Terminal

Product terminals typically are separate from an oil and gas processing plant but in some cases may be associated with a refinery. The product typically handled at a terminal includes gasoline, diesel, fuel oil, liquefied petroleum gas (LPG), kerosene, aviation gasoline and jet fuels. The few environmental concerns encountered in product terminals are similar to those in an oil and gas processing plant and the pollution control methods for a product terminals are similar to an oil and gas processing plant [106–109].

Some other water pollution sources include Hydraulic fracturing which is a practice that may involve the injection of known toxic chemicals into or close to water supplies. Storm water runoff during construction or runoff from established well pads can introduce sediment and toxic chemicals into nearby rivers and streams. Storage and disposal of drilling and production wastes in pits can contaminate groundwater and surface waters.

4.2 Design Procedure for Effluent Water Pollution Control

Water pollution can be controlled in the multiple ways. It is best controlled by the dilution of water. The pollutants must be treated chemically and must be converted into the non-toxic substances.

Examples of design and procedures which are generally beneficial are as follows:

- Recovery of oil spills and hydrocarbons with vacuum trucks to reduce emissions and water effluents [5–8, 110–113].
- A specialized program for handling oily wastes, sludge, wash waters and other effluents.
- Maximization of air fan cooling and employ cooling water only for those services in which low process temperatures make air fan cooling impractical or uneconomic.
- Separation of oily wastes, concentrated wastes and other process wastes from general effluents for more effective treatment.
- Limiting the amount of water used for process unit wash downs.
- Converting foul water strippers to reboiler strippers to reduce foul water and recover condensate.
- Reduction in shock pollutant loads on treatment facilities through the periodic flushing of process sewers to prevent contaminant buildup and by the use of flow and load equalization prior to treatment.
- Using caustic injection into desalted crude to reduce NH_3 needed to control corrosion in the crude unit overhead system.

4.3 Spill Prevention and Control

Among specific design parameters are impervious dikes around tankage (feedstock and product), containment of storm water from the process area(s), ability to treat contaminated storm water in the wastewater treatment facility, leak detection systems capable of detecting small volume or slow rates of leakage from the pipeline system and appropriate use of valves to minimize potential spill volumes [114–117].

4.3.1 Spill Prevention Techniques

A key component of an integrated approach to pollution prevention is to minimize accidental and incidental releases of toxic and hazardous materials to the atmosphere. These releases usually result in not only a waste of material, but also in the generation of contaminated soil, absorbent material and contaminated product that has to be treated and disposed. A structured plan is absolutely necessary to ensure control of systems and verify that the goal of zero spills can be achieved. Prevention of spills is the first line of defense in protecting life, property and the environment. Experience has shown that operational or human error and equipment failures are the principal causes of spills. Both can be reduced through the involvement and commitment of all staff to spill prevention.

Proper design, inspection and maintenance of general facilities are of principle importance. Operator capacity is also extremely important and must be periodically tested and upgraded [117–120].

Given good equipment, good operators and good procedures, spills will be reduced. They will not, however, be eliminated. The difference between a minor event and a catastrophic event depends almost entirely on planning. Such planning includes plant design with spill containment features, and alarms, a workable and efficient contingency plan, trained spill control personnel, and adequate spill control equipment.

4.3.2 Bulk Storage

It is a requirement that onshore oil production or bulk storage facilities provide oil spill prevention, preparedness and responses to prevent oil discharges. Oil storage tank construction and material should be compatible with the oil stored and the storage conditions such as pressure, temperature, etc.

Impervious secondary containment should be provided for the capacity of the largest single tank plus a sufficient allowance for precipitation and free board.

New metallic tanks buried underground should be protected from corrosion by coatings, cathodic protection, or other effective methods compatible with local soil conditions. The use of non-metallic tanks, if available and practical, should be given consideration.

Plant effluents which are discharged into a watercourse should have disposal facilities observed frequently enough to detect any possible system upset that could cause an oil discharge.

Above-ground tanks should be subjected to appropriate integrity testing. Appropriate procedures might include hydrostatic testing, visual inspections, or inspection by a system of non-destructive shell thickness gauging [116–122].

4.4 Groundwater Pollution Control

4.4.1 Basic Sources

Obviously, contaminated groundwater is very difficult and expensive to clean up. Solutions can be found after groundwater has been contaminated but this is not always easy. The best thing to do is adopt pollution prevention and conservation practices in order to protect important groundwater supplies from being contaminated or depleted in the first place.

Two basic sources of spilled liquid petroleum products are equipment failure and operator error. Equipment failure includes corrosion and leaking of both above- and below-ground piping and tanks, valves failure, refinery unit upsets, and sewer and drain leaks. Many of these failures may be avoided through proper inspection and maintenance procedures.

Operator error includes overfilling tanks and improper alignment of valves and piping. These and other operator errors can best be corrected through developing proven operating procedures, regular training and testing of personnel, and systematic follow-up to assure that procedures are followed.

4.4.2 Preventive Measures

Many steps are being taken to keep pollutants from reaching groundwater supplies. The spill prevention control and countermeasure (SPCC) regulations strive to prevent oil from entering navigable waters through the prevention, control and mitigation of oil spills. The preventive measures to be installed during the construction of a permanent structure must consider the following:

- The type of construction (refinery, storage tank, pipeline, etc).
- The volume and the nature of the oil likely to pollute the site.
- The geology and hydrogeological environment: nature of the terrain, depth, activity and quality of the aquifer.
- The economic environment: proximity to and capacity of water wells and intakes for domestic purposes, risk of pollution of a river, etc.
- The preventive system involves four areas: corrosion protection, surface preventive measures, subsurface preventive measures and monitoring devices to detect and warn of unsuspected pollution not visible from the surface or of a dangerous change in groundwater levels.

Other methods of preventing spilled product from entering the ground and controlling its direction are as follows:

- Rendering the soil impermeable where required by means of a concrete paving, a clay or bitumen layer, plastic sheets (PVC sheets covered with gravel, fiber-glass-reinforced epoxy) and chemical to be mixed with the soil.

- A surface drain system in the plant area carrying all oil and oil-contaminated water to a dirty water sewer, and then to an interceptor or separator, by means of a pipe system with manholes (cast iron, steel, epoxy) and gutters [118–124].

4.4.3 Types of Devices

4.4.3.1 Trench

This system of protection, which is used as a barrier to prevent the horizontal movement of the oil, can only be carried out on a practical scale if the water table is situated at a depth of less than about 3–8 m depending on soil conditions.

Spread of oil on the groundwater surface is intercepted by digging the trench to about one meter below the piezometric level. Oil flows onto the water surface where it can be recovered.

4.4.3.2 Hydrodynamic Protection

The principle of oil spill control using hydrodynamic methods is to effect a change in the groundwater flow pattern such that the free oil or the contaminated water, as the case may be, can be drawn to a specific control point or points. This can be achieved by discharging or recharging the aquifer, or a combination of both. The success of the method depends on maintaining an artificial gradient in the groundwater surface.

4.4.3.3 Monitoring

Groundwater monitoring devices are typically installed to detect and warn of unsuspected contamination not visible from the surface or of a dangerous change in groundwater levels. These devices are installed around petroleum storage areas, waste treatment/disposal facilities (including lagoons, land farms and landfills), or an entire facility depending on the potential for contamination.

Care should be taken in choosing monitoring devices to maximize accuracy and reliability of the system. Also, monitoring should be conducted to differentiate between previous and new spills.

4.4.3.4 Mitigation Measures

After a spill or any contamination is detected, remedial measures such as determination of the extent or contamination (boundaries) and a hydrogeological assessment of the contamination area to determine the corrective action necessary must be conducted. When an appropriate action is determined, steps including recovery of the oil and oily water and restoration of the site may be instituted.

4.4.3.5 Recovery Measures

The principal factor to be considered when recovering free oil from the groundwater surface is to employ the natural water gradient or by inducing one or by increasing an existing gradient by artificial means. By good use and operation of recovery equipment, free oil can be concentrated at a relatively limited number of selected sites and removed. Wells and trenches are typically used for recovering oil and water.

4.5 Wastewater Pollution Control

A major environmental concern with terminal operations is oil being spilled and the effects on birds and marine life. Depending upon the type of terminal (offshore or onshore) and the characteristics of the water (such as currents and proximity to open water), the effects of a spill can vary from insignificant to extremely damaging. For example, an enclosed area such as an inlet from the sea, which has been described as the most productive marine environment, may experience accumulation of oil to unacceptable levels over time.

The largest source of contaminated water and wastewater at a crude terminal is the ballast water from tankers. The quantity of ballast water requiring treatment depends upon the ship design, operation and regulations governing the discharge of ballast waters. The ship-design parameters include the amount of segregated ballast, the tank dimension, and the use of onboard oil/water separators. The operating parameters include the type of previous cargo, weather conditions and tank-cleaning procedures. Optimizing the design and operation of a tanker can reduce the amount of water requiring treatment.

Ballast water treatment has consisted of settling the ballast in shore side tankage for periods of 10–24 h, skimming off oil and discharging the water. This simple gravity separation may still be acceptable in some circumstances. For a better quality effluent, physical, chemical and/or biological methods are necessary.

In some locations where shore space is at a premium offshore, deballasting facilities in the form of converted redundant tankers are utilized. One method to eliminate contaminated ballast water in tankers is the use of segregated ballast [115–119].

These treatment systems rely on gravity difference to separate the oil and water. They are capable of removing the bulk of non-dissolved and non-emulsified oil. Examples include storage and settlement, once-through storage with skimming, gravity separators, corrugated plate interceptors (CPI) and holding basins.

Various combinations of the above individual treatment steps are used and such combinations may consist of storage and settlement plus gravity separators or CPI; storage and settlement plus holding basin; or storage and settlement plus gravity or CPI plus holding basin.

There are several reasons why a combination of steps is often the best choice of treatment process. First, storage and settlement ahead of a CPI or gravity separator will remove crude oil and prevent temporary overloading of the downstream separator.

Second, rather than sizing the CPI or gravity separator to handle the maximum hourly flow rate, costs can be reduced by having a combination of a settling tank and CPI or gravity separator, with the tank designed for the maximum flow rate and the separator design to deal with the average flow rate. Third, a CPI or separator plus holding basin serves as a guard chamber to finally trap any inadvertent discharge of oil from the settling tank.

Oil spills often result in both immediate and long-term environmental damage. Some of the environmental damage caused by an oil spill can last for decades after the spill occurs.

4.5.1 Biological Treatment

The idea behind all biological methods of water treatment is to introduce contact with bacteria (cells), which feed on the organic materials in the wastewater, thereby reducing its biochemical oxygen demand (BOD) content. In other words, the purpose of biological treatment is BOD reduction. Biological treatment may in some cases be appropriate for removing dissolved biodegradable materials, which are often in low concentrations in normal ballast water. Typical devices used for biological treatment include activated sludge, trickling filters, rotating disks and lagoons (aerated or not).

4.5.2 Spills

An oil spill is the release of a liquid petroleum hydrocarbon into the environment, especially marine areas, due to oil and gas industries activity, and it is a form of pollution. A major environmental concern with terminal operations is oil being spilled and the effects on birds and marine life.

Depending upon the type of terminal (offshore or onshore) and the characteristics of the water (such as currents and proximity to open water), the effects of a spill can vary from insignificant to extremely damaging. For example, an enclosed area such as an inlet from the sea, which has been described as the most productive marine environment, may experience accumulation of oil to unacceptable levels over time.

The oil that is spilled offshore will have less impact on the marine environment than an equivalent oil spill within an inlet from the sea (estuary) for three reasons:

- There will be fewer organisms offshore to be affected;
- The concentration of toxic compounds within the water column is expected to be less because more dilution water is available in offshore and;
- The contact time with the marine life will generally be shorter for an offshore spill due to the restricted flushing of the inlet from the sea.

In addition to spilled oil, the treated ballast water can also affect the marine life in an "enclosed" area. Thus, it may sometimes be best to pipe the treated water to another location where good mixing may occur, thereby, protecting the "enclosed" area and minimizing the effects on the marine environment.

Oil spilled by damaged tankers, pipelines or offshore oil rigs coats everything it touches and becomes an unwelcome but long-term part of every ecosystem it enters. When an oil slick from a large oil spill reaches the beach, the oil coats and clings to every rock and grain of sand. If the oil washes into coastal marshes, mangrove forests or other wetlands, fibrous plants and grasses absorb the oil, which can damage the plants and make the whole area unsuitable for wildlife habitat.

When some of the oil eventually stops floating on the surface of the water and begins to sink into the marine environment, it can have the same kind of damaging effects on fragile underwater ecosystems, killing or contaminating many fish and smaller organisms that have essential links in the global food chain.

Spill contaminant is an important feature of a terminal. The most common types of barrier systems used for floating booms and sorbent ropes. Two less frequently used alternatives are the air-bubble barrier systems and an enclosed berth [120–125].

4.5.3 Residual Suspended Matter

Some pollutants do not dissolve in water as their molecules are too big to mix between the water molecules. This material is called particulate matter and can often be a cause of water pollution.

- The suspended particles eventually settle and cause thick silt at the bottom. This is harmful to marine life that lives on the floor of rivers or lakes.
- Biodegradable substances are often suspended in water and can cause problems by increasing the amount of anaerobic microorganisms present.
- Toxic chemicals suspended in water can be harmful to the development and survival of aquatic life.

A major design requirement for an oil and gas facility is to minimize or eliminate the emission of pollutants to the environment. The methods necessary to implement this depend on the types of crude oil processed, the types of products, the availability of water fuel and other utilities and the agreed pollution parameters. Safety requirements must also be considered at this stage to ensure that the surrounding population and plant personnel are protected from hazards such as fire, explosion and toxic chemicals.

If it is necessary to reduce the non-dissolved oil content of the effluent below 25 mg/l, there are several processes available which can be added after the chosen simple gravity system.

Such methods will also reduce suspended solids to below approximately 30 mg/l. BOD can also be reduced because of the oil and solids removed. These processes have little effect on soluble oil content.

The physical methods include dissolved air flotation, filtration (using gravity or pressure filters), physical separation plus use of chemicals such as inorganic flocculants, and/or demulsifiers or poly electrolytes, flocculation/sedimentation, flocculation-dissolved air flotation and induced air flotation.

In the following pages, the general outline of pollutant specific for management in the siting and design of oil, gas and chemical processing plant or oil terminal will be discussed.

The items of concern in siting and design of a facility are briefly reviewed. They are meant only to point out potential environmental effects.

4.6 Design Aspects

4.6.1 Aquatic Ecosystems

An aquatic ecosystem is an ecosystem in a body of water. Communities of organisms that are dependent on each other and on their environment live in aquatic ecosystems. Wetlands, rivers, lakes and coastal estuaries are all aquatic ecosystems—critical elements of Earth's dynamic processes and essential to human economies and health. An oil, gas and chemical processing plant should take into account potential impacts to aquatic ecosystems. The characterization of aquatic systems in a siting should include the location of spawning areas, feeding zones, commercial fishing areas, and sport fishing areas, a description of the benthic populations, and an estimate of primary productivity and its limiting factors.

Design considerations should include the facility's water supply requirements. While water use is usually non-consumptive, attention should be given to the intake and discharge zones.

At refineries or any other oil and gas facilities, there is always the potential for a release of oil or petroleum products. All facilities should have a spill contingency plan and the basic pieces of equipment necessary to cleanup a spill. All tankage should be properly diked [122–127].

4.6.2 Terrestrial Ecosystems

Terrestrial ecosystems are distinguished from aquatic ecosystems by the lower availability of water and the consequent importance of water as a limiting factor. Terrestrial ecosystems are characterized by greater temperature fluctuations on both a diurnal and seasonal basis than occur in aquatic ecosystems in similar climates. The availability of light is greater in terrestrial ecosystems than in aquatic ecosystems because the atmosphere is more transparent than water. Gases are more available in terrestrial ecosystems than in aquatic ecosystems. Those gases include carbon dioxide that serves as a substrate for photosynthesis, oxygen that serves as a substrate in aerobic respiration, and nitrogen that serves as a substrate for nitrogen fixation. Impacts to terrestrial ecosystems from the siting of a refinery or terminal include a reduction or loss in total available habitat, destruction, or modification of food webs, and changes in populations. Another concern in the siting should be the sensitivity of plants and animals to pollutants.

4.6.3 Wetland Ecosystems

A wetland is a land area that is saturated with water, either permanently or seasonally, such that it takes on the characteristics of a distinct ecosystem. Primarily, the factor that distinguishes wetlands from other land forms or water bodies is the characteristic vegetation that is adapted to its unique soil conditions: Wetlands consist primarily of hydric soil, which supports aquatic plants.

Where the water table is at, near or above the land surface for a significant part of most years, the hydrologic regime is such that aquatic or hydrophytic vegetation is usually established, although tidal flats may be non-vegetated. Because wetlands are water systems, any alteration affecting the movement of quality of water in a small area is transmitted to other areas magnifying potential impacts.

Wetlands may be altered directly by filling, dredging, draining or creating impoundments. Indirectly, alteration of water flow patterns at the locations and changes in adjacent land use can change the functions and values of wetland areas.

In addition to facility construction, the laying of pipelines associated with these facilities can have impacts on wetlands.

4.6.4 Water Pollution Control

Pollution is the introduction of contaminants into an environment that causes instability, disorder, harm or discomfort to the ecosystem. The Environmental Pollution Control identifies and examines characteristics and sources of the major environmental pollutants. Oil and gas operations have a variety of impacts in the environment through the pollution. These impacts depend upon the nature, stage, complexity and size of the operation and sensitivity of the surrounding environment. The major environmental concerns associated with oil and gas operations are oil spills, drilling waste fluids or mud, drilling waste solids, produced water, and volatile organics.

The siting and design guidelines for water pollution control are as follows:

The most important aspect of water pollution control in the siting of an oil and gas processing plant is the effect of the wastewater effluent on the receiving water. Several factors which should be assessed in the siting investigation are heat load, total dissolved solids, heavy metal concentrations and the effects of organics in the effluent.

The design of a water pollution control depends on the degree of cleanup required to permit discharge of the wastewaters to either a body of water or public wastewater treatment plant and by the characteristics of the oil, gas or chemical processing plants. The following design practices are directed primarily toward segregation of process and non-process wastewaters and the recycling and reuse of raw and treated wastewaters:

- Oil and gas facilities wastewaters should be segregated and treated based on oil content and potential for reuse. Four common divisions are oil-free wastewaters, oily cooling water, process water and sanity wastewaters.
- Raw and treated wastewater streams should be recycled to reduce the effluent volume and thereby the makeup water required. Specific practices include the following:

  - use of catalytic cracker accumulator wastewaters rich in H_2S (sour waters) for make-up to crude desalters after stripping;
  - use of blowdown condensate from high-pressure boilers for make-up to low-pressure boilers;
  - reuse of waters that have been treated for closed cooling systems, fire mains and everyday washing operations;
  - stormwater use for routine water applications;
  - use of blowdown waters from cooling towers for water seals on high-temperature pumps;
  - recycling of steam condensate;
  - recycling of cooling waters;
  - pipe still overhead waters can be sent directly to the desalter.

- Water conservation practices should be employed to reduce the volume of wastewater requiring treatment. Among the most common practices are as follows:

  - recycle and reuse of refinery wastewaters;
  - increase the use of air cooling;
  - replacement of barometric condensers with surface condensers and vacuum pumps;
  - process modifications that have reduced water requirements;
  - use of closed pump gland cooling water systems.

- Where possible, treat highly contaminated streams at their source.
- Eliminate waste products in the process operations before they become associated with waste streams.
- Segregate wastewaters for treatment according to the degree required for reuse, recycle or discharge.
- Design sample connections to eliminate prolonged purging.
- Design facilities to avoid all spills.

Design the treated wastewater effluent outfall to minimize any environmental damage. For example, discharge of effluent into a salt marsh may cause severe damage; therefore, the discharge should be piped to an offshore location [123–139].

4.6.5 Washing Water and Process Water

Considering the properties of fouled water, washing water is identical to process water, so the former is frequently called "process water".

Water is used in washing because it dissolves various substances. For example, carbon dioxide, hydrogen sulfide, hydrochloric acid in gas are dissolved in dilute alkali water.

The amounts of washing water and process water being used in oil, gas and petrochemical plants are small compared with those of cooling water. These wastewaters contain a considerable amount of organic substances and dissolved oils. But the COD and BOD values are insufficient to indicate the amounts of organic substance contained in the wastewater from the relevant oil, gas and chemical processes. It is necessary that the organic matter in wastewater from petrochemical plants be indicated using total organic carbon (TOC) and total oxygen demand (TOD) but the TOC and TOD indications themselves are not always considered useful as criteria for the detection of toxic substances in the wastewater or for the selection of wastewater treating methods. As organic compounds discharged from petrochemical processes have a variety of chemical properties depending on the processes, it is necessary to pursue the emission sources [125–130] (Table 4.1).

4.6.6 Typical Pollutants of Petrochemical Industry

Aqueous effluents originating from the petrochemical industry are essentially characterized by the presence of the following substances:

- Organic substances that are not biodegradable, or only slightly so.
- Nitrogen compounds.
- Heavy metals.

Table 4.1 Type of water pollution

Type of operation	Examples of pollution
Water pollution (Wastewater, waste liquid)	
Cooling water	Direct cooling or quenching of decomposed gas—Wastewater contains tar dust, hydrogen sulfide and cyanide
	Breakage of indirect cooler tubes-contamination of cooling water due to liquid inside the tubes
	Cooling water of pumps, etc.—water is contaminated by oil, etc.
Boiler feed water	Steam ejector—steam condensate from ejector contains volatile hydrocarbon
Washing water	Water containing hydrogen sulfide and hydrochloric acid, etc. is discharged from the washing of gas
	Water containing hydrochloric acid, etc. is discharged from the washing of liquid
	Dust is contained in discharged water from dust collector
Process water, feed water (chemical reaction, electrolysis)	Solvent for suspension and emulsion polymerization contains catalysts, emulsifier, plastic monomer, etc.
	Steam condensate from steam stripping—dissolve hydrocarbons
	Steam condensate originating from dilution steam for naphtha thermal cracking—contain carbon, phenol and light oil
Leakage (loss)	Leakage from pump and agitator shafts, valve stems, and flanges, and due to operational error

4.6.7 Chemical Waste Treatment

Chemical waste is generally treated by waste disposal companies according to its physical and chemical characteristics. For example, Acetone, a flammable liquid has properties that make it a fuel as well as a solvent and hence, when it is treated as a waste stream, it is blended with other flammable chemicals and its energy value is then recovered as a fuel. Toxics are generally diluted and fixed within a block of concrete material and are in essence immobilized so that there is no leaching to the environment.

Petrochemicals have been defined as bulk chemicals derived principally from natural gas, petroleum or both. A careful check should be made of processes proposed or used for the manufacture of petrochemicals to decrease the possibility of water soluble organics entering water supplies. The following methods should be considered:

- Recycling and reuse of waste streams.
- Quenching with oil or chemicals other than water that do not produce waterborne wastes.
- Use of alternative processes that do not produce waterborne wastes.

- Use of air coolers or of cooling towers in place of once-through cooling water.
- Elimination of waste products in the manufacturing operation before they become associated with waste streams.
- Processing of waste streams to reduce the amount of chemicals in wastewaters leaving the plant.

The extensive use of automated controls, alarms and checks by the operators to prevent loss of chemicals is also important. It is essential that adequate facilities be installed to prevent uncontrolled release of chemicals and wastes to sewers or receiving waters. A very effective means of quality control is the use of large lagoons capable of holding several days' production of wastewater; this allows the water to be checked before being released to receiving waters [131–135].

4.7 Fertilizer

4.7.1 General Appraisal

The liquid effluents arising from a fertilizer factory originate from a variety of sources and may be summarized as follows:

- Ammonia-bearing wastes from ammonia plants.
- Ammonia and urea wastes from urea-manufacturing plant.
- Ammonium salts such as ammonium nitrate, ammonium sulfate and ammonium phosphate.
- Phosphates and fluoride wastes from phosphate and super phosphate plants.
- Acidic spillages from sulfuric acid, nitric acid and phosphoric acid plants.
- Spent solutions from the regeneration of ion exchange units, acid from cation exchange units and alkali from regeneration of anion exchange units.
- Phosphate, chromate, copper sulfate and zinc wastes from cooling tower blowdown.
- Salt of metals such as iron, copper, manganese, molybdenum and cobalt.
- Sludge discharged from clarifiers and backwash water from sand filters.
- Carbon slurry from partial oxidation units.
- Scrubber wastes from gas purification processes containing contaminants such as:

  - Mono and di-ethanol amines (MEA and DEA).
  - Arsenic as As_2O_3.
  - Potassium carbonate.
  - Caustic soda.

The effluents from fertilizer manufacturing are generated from a wide range of unit operations, and considerable variation between wastewaters from different factories may be noted. The age, state of repair, operational management and degree of sophistication of each manufacturing unit will play an important role in determining the degree of in-plant materials loss, and the important factors leading to excessive losses (and subsequent pollution) may be summarized as follows:

- Outdated basic manufacturing plant with low efficiency and poor process control.
- Improper maintenance and repair, with particular emphasis on servicing of control equipment.
- Variations in feedstock and difficulties in adjusting process plant to cope with these variations effectively.
- Lack of consideration for pollution abatement and the prevention of materials loss at the original plant design stage.

The overall water requirement for fertilizer manufacturing plants may be high, due to process cooling requirements. The total volume of effluent discharged is dependent to a considerable extent on the degree of in-plant recirculation, and in the case of total recycle, the raw water is used primarily for make-up purposes. Plants designed on a once-through-process cooling system generally give rise to high volumes of effluent, from $1,000 \text{ m}^3 \text{ h}^{-1}$ to volumes in excess of $10,000 \text{ m}^3 \text{ h}^{-1}$, consisting primarily of cooling water discharge [122–133].

4.7.2 Nitrogenous Fertilizers

A complex nitrogenous fertilizer plant based on the production of ammonium nitrate and urea products may give rise to pollutants such as ammonium nitrate, nitric acid, ammonia, urea, sulfuric acid, caustic soda, chromate, oil, grease and boiler feed additives, contained within the overall effluent stream. The individual effluents from the ammonia and urea plants could be categorized as follows:

(a) **Ammonia plant**

- HCN stripper outlet.
- Catalyst reduction.
- Shift process condensate.

(b) **Urea plant**

- Concentrate liquor.
- Cooling water blowdown.

Additional pollutant discharges may arise from oily water and sanitary sewage effluents.

4.8   Effect of Pollution

Wastes may be subcategorized into major and minor elements, but it should be noted that in specific instances, particular minor waste components may exercise significant polluting effects.

4.8.1 Major Pollutants

General water pollution effects from fertilizer manufacturing wastes are dependent primarily on the elements nitrogen and phosphorus, in their varying chemical forms.

4.8.2 Ammoniacal Nitrogen and Urea

These two compounds are grouped together since urea may be hydrolyzed to ammoniacal nitrogen. The pollution problems which may be attributable to ammoniacal nitrogen include toxicity, oxygen demand and eutrophication. Ammonia can be toxic to fish and other aquatic life forms at relatively low concentrations, and urea itself may be toxic to some aquatic life.

Ammoniacal nitrogen and urea may both be oxidized biologically. As such, their presence must be considered a potential oxygen demand in receiving water. In addition, ammoniacal nitrogen may act in its role as a fertilizer in an aquatic environment, leading to excess growth of algae and aquatic macrophytes and contributing toward accelerated eutrophication.

The presence of high ammoniacal levels may also cause problems if the receiving water is used for water supply purposes, due to chemical interference with chlorination (i.e., formation of chloramine intermediates) and resultant increase in chlorine demand.

4.8.3 Nitrate

The water pollution problems resulting from high nitrate levels may be categorized into eutrophication and public health effects. High levels of nitrate can give rise to increased eutrophication, leading to the promotion of growth of algae and macrophytes, adversely affecting water quality and amenity value. Health hazards related to nitrate in water used for supply purposes are considered to be infant methemoglobinemia and carcinogenic potential.

4.8.4 Phosphate

The presence of significant levels of phosphate has important effects on eutrophication. In terms of inorganic nutrient enrichment of receiving waters, phosphate may in many instances be more important than nitrogenous compounds, due to the fact that some forms of aquatic plant life may fix atmospheric nitrogen, so removing the absolute requirement for soluble forms of nitrogen to promote growth.

Under these circumstances, phosphate becomes the growth-limiting agent, and programmes to control eutrophication have generally sought to reduce available phosphate limits, to prevent excessive algal and macrophyte growth, with subsequent increase in nutrient retention.

4.8.5 Minor Constituents

In addition to pollution arising from the discharge of nitrogenous or phosphatic elements in liquid waste streams, pollution can be caused by a number of secondary waste components. The most important of these may be listed as follows:

(a) Oil and grease
(b) Hexavalent chromium
(c) Arsenic
(d) Fluoride

In specific instances, one or more of these individual pollutants may give rise to detrimental effects in a receiving water, due primarily to toxicity, or can cause inhibition of nitrification. In addition, oil and grease may adversely affect the oxygen transfer characteristics of a watercourse.

4.9 Water Monitoring

Monitoring may be necessary for a number of reasons, such as:

(a) To measure water problems (what, where, why, when).
(b) To measure waste parameters for use in calculation of waste treatment charges for municipal or regional treatment system.
(c) To measure waste parameters to allow detection and triggering of emergency actions in case of spills or process upsets.
(d) To measure the effect of a wastewater discharge on the quality of a receiving body of water.
(e) To measure the quantity and quality of all process wastes as well as influent liquid raw materials.
(f) To gather enough information on water used and contamination to allow design of pre-treatment and/or reclamation systems.

LOGP Industries discharging to municipal or regional treatment systems shall be required to monitor their wastewater discharges or allow monitoring by others [120–137].

4.10 Coal Bed Methane Produced Water

Coal bed methane (CBM) gas recovery techniques are unique compared to other production methods. Formation water must be removed, or "dewatered" as it holds the methane gas in the coal seam by hydrostatic pressure. Removing the formation water depressurizes the formation, thus releasing the gas to production. Initial water volumes are very high, but decrease rapidly to allow for the release of the methane gas. Producers must manage these considerable volumes of water generated during the dewatering process. Much of the water can be disposed of by direct discharge given the high quality of the CBM produced water in a coal seam gas basin. Produced water of a lower quality, however, must be managed depending on environmental compliance and economic objectives. This would include volume of produced water, proximity to surface water, rights-of-way, influent chemistry, discharge quality requirements, land use provisions (public or private) and recycle objectives.

Designs incorporate reverse osmosis (RO) and recovery reverse osmosis (RRO), as this configuration has proven effective for meeting produced water treatment objectives. The RO/ RRO process shall be permitted through an environmental quality/water quality division (WDEQ). Both plants minimize waste by maximizing system recovery and use an aeration pond for evaporating and concentrating the brine. The plants should be designed with bypass and blend provisions, so the produced water can be blended to a wide range of discharge specifications. Both plants maximize membrane performance with filtration and scale control, but differ in the approach to controlling scale. It is the nature of the scale control [137–139].

4.10.1 Influent and Effluent Criteria

Feed water characteristics must be clearly understood in order to properly design the treatment plant. This includes seasonal variability that may identify influent extremes or complex chemistries. Waste and product stream characteristics must also be understood so that service factor, redundancy and compliance can be addressed in the plant system design. By its inherent nature, CBM water is high in sodium and bicarbonate and low in hardness and may also include suspended solids, iron, silica and barium. Sodium is a closely monitored aspect of the treatment plant effluent. Soils with an excess of sodium ions, as compared to calcium and magnesium ions, can impact the way plants adsorb water. The ratio of the sodium to calcium and magnesium is referred to as the sodium adsorption ratio (SAR).

Application engineers use solubility indices to understand the relationship of the dissolved ions as they move through the treatment process. For instance, one technique for predicting calcium carbonate solubility considers the bicarbonate/carbonate and calcium concentration to access the potential for hardness scale formation. This is the concept behind the Langelier Saturation Index (LSI). A positive LSI denotes an increased potential for calcium carbonate scale formation while a negative LSI denotes that calcium carbonate may dissolve in the solution. LSI is one of the many solubility indices that facilitate designer engineers' understanding of ion interactions as water chemistries change through a process. This information helps designers control the severity of the process and applies appropriate equipment and chemistries to moderate the behavior of the water as it progresses through the process.

Another constituent common in CBM water is silica. Because of its unique chemistry, silica poses special treatment challenges to design engineers. While the silica concentration in Powder River Basin—produced water is moderate, the high recovery rate of the membrane system creates ideal conditions for silica to scale membrane surfaces. Silica precipitation control is further complicated, since control techniques for other ions conflict with methods for controlling silica [138–140].

4.10.2 Geographical and Environmental Concerns

This must be considered when designing the system to ensure sufficient redundancy to address uptime and reliability objectives. Key considerations include redundancy, call-out features, response time and safety.

Acid feed systems must be carefully designed to minimize risks to personnel and facility. The volume of hydrochloric acid needed to neutralize the alkalinity inherent in the CBM water is considerable. The acid is delivered in tanker trucks, often down lease roads and potentially during severe weather. The acid should be stored outdoors in double-contained tanks. Feed lines and valves should also be double contained. Tanks should be located as close as possible to injection points to minimize the length of the feed lines.

Another key criterion in system design is meeting discharge specifications to comply with standard specifications for protecting aquatic life from toxicity. The test commonly used to confirm compliance is referred to as Whole Effluent Toxicity Test, or WET test. Effluent samples are collected at appropriate outfalls and analyzed to determine the impact of the discharge water on aquatic life in the receiving waters. Toxicity occurs if mortality exceeds 50 % for either species at the effluent concentrations. The subject water is diluted with synthetic lab water to evaluate the degree of toxicity as compared to the laboratory control sample [137–140].

Chapter 5
Fire Water Storage Facilities and Distribution

Keywords Fire-fighting water system · Pumping facilities · Winterizing · Fire protection · Water spray–fixed systems · Fire extinguishment · Explosion prevention

5.1 Introduction

Water is the most common fire extinguishing agent used due to its abundance, low cost and effectiveness. It is the most commonly used agent for controlling and fighting fire, by cooling adjacent equipment and for controlling and/or extinguishing the fire either by itself or combined as foam. It can also provide protection for fire fighters and other personnel in the event of fire. Water must therefore be readily available at all the appropriate locations, at the correct pressure and in the required quantity. Defining what constitutes an adequate supply of water for fire fighting is also central to planning fire service operations.

Fire water should not be used for any other purpose. Unless otherwise specified or agreed, the main requirements are given for major installations such as refineries, petrochemical works, crude oil production areas where large facilities are provided and for major storage areas.

This chapter specifies minimum requirements for water supply for fire-fighting purposes. It is important that engineers and practitioners provide and maintain these minimum water supplies and discussions with municipality fire stations that would not only include the water available from the hydrants but also help to assure the continuous and adequate flow of water for fire fighting in oil, gas and chemical processing industries [141–146].

In determining the quantity of fire water, that is, required fire water rate, protection of the following areas must also be considered:

- General process;
- Storage (low pressure), including pump stations, manifolds and in line blenders;
- Pressure storage (LPG etc.);
- Refrigerated storage (LNG etc.).

A. Bahadori et al., *Essentials of Water Systems Design in the Oil, Gas, and Chemical Processing Industries*, SpringerBriefs in Applied Sciences and Technology, DOI: 10.1007/978-1-4614-6516-4_5, © The Author(s) 2013

The establishment of fire-fighting water requirements is central to fire service operations as it underpins the selection and distribution of resources. Basically, the requirements consist of an independent fire grid main or ring main fed by permanently installed fire pumps taking suction from a suitable large capacity source of water such as storage tank, cooling tower basin, river, sea. The actual source will depend on local conditions. The amount of fire-fighting water needs to be specified by pressure, flow rate and total available quantity. The provision of sufficient fire-fighting water is to ensure that the fire service can curtail and suppress a fire.

The water will be used for direct application to fires and for the cooling of equipment. It will also be used for the production of foam [142–149].

5.2 Public Water Systems

Adequate water supply is critical for effective fire fighting. Where a non-reticulated water supply is provided or the reticulated water supply is deemed inadequate, an additional onsite-stored supply of water for fire fighting will be required.

Adequacy of water supply needs to be determined by flow tests or other reliable means. Where flow tests are made, the flow in (liter/minute) together with the static and residual pressures needs to be indicated on the plan.

One or more connections from a reliable public water system of good pressure and adequate capacity furnish a satisfactory supply. A high static water pressure is not, however, be the criterion by which the efficiency of the supply is determined.

If this cannot be done, the post-indicator valves need to be placed where they will be readily accessible in case of fire and not liable to injury. Where post-indicator valves cannot readily be used, as in a city block, underground valves must conform to these provisions and their locations and direction of turning to open need to be clearly marked.

Where connections are made from public waterworks systems, it may be necessary to guard against possible contamination of the public supply. The requirements of the public health authority must be determined and followed [144–148].

5.3 Bases for a Fire-Fighting Water System

If fire fighters are unable to maintain an uninterrupted supply of water on the fire, the result can be a relatively unchecked spread of the fire, leading to the complete loss of structures or an extension of the fire beyond the capabilities of the emergency personnel involved.

The water supply must be obtained from at least two centrifugal pumps of which one is electric-motor-driven and one driven by a fully independent power source, for example, a diesel.

The provision of fire-fighting water requires consideration of a number of points. These are as follows:

- Cost;
- Reliability;
- Quality of water;
- Water demand (i.e., flow rate, storage and available pressure);
- Provision for access via fire hydrants.

A ring main system must be laid around processing areas or parts thereof, utility areas, loading and filling facilities, tank farms and buildings while one single line must be provided for jetties and a fire-fighting training ground complete with block valves and hydrants.

The water quantities required are based on the following considerations:

- There will be only one major fire at a time.
- In processing units, the minimum water quantity is 200 dm^3/s or air foam making and exposure protection. It is assumed that approximately 30 % of this quantity is blown away and evaporates; the balance of this quantity, which is 140 dm^3/s per processing unit, must be drained via a drainage system.

The quantity of fire water required for a particular installation should be assessed in relation to fire incidents which could occur on that particular site, taking into account the fire hazard, the size, duties and location of towers, vessels, etc. The fire water quantity for installations having a high-potential fire hazard should normally be not less than 820 m^3/h and not greater than 1,360 m^3/h.

- For pressure storage areas, the quantity needed for exposure protection of spheres by means of sprinklers.
- For jetties, the quantity needed for fighting fires on jetty decks and ship manifolds with air foam as well as for exposure protection in these areas.
- For storage areas, the quantity needed for making air foam for extinguishing the largest cone roof tank on fire and for exposure protection of adjacent tanks.
- The policy for a single major fire or more to occur simultaneously must be decided upon by the authorities concerned.

For new installations the quantities required for items mentioned above must be compared, and the largest figure must be adhered to for the design of the fire-fighting system.

The system pressure must be such that at the most remote location, a pressure of 10 bar can be maintained during a water takeoff required at that location.

Fire-fighting water lines must be provided with permanent hydrants. Hydrants with four outlets must be located around processing units, loading facilities, storage facilities for flammable liquids and on jetty heads and berths. The amount of fire-fighting water needs to be specified by pressure, flow rate and total available quantity.

Hydrants with two outlets must be located around other areas, including jetty approaches. Although water is inexpensive and readily available, its processing and distribution carry a significant cost in terms of infrastructure cost [150–156].

5.4 Fire Water Pumping Facilities

Fire water pumps are one of the most responsive and reliable fire-fighting systems available. High-reliability fire pumps are of critical importance and must always be ready for rapid response for extinguishing fires in oil and gas facilities. Fire water must be provided by at least two identical pumps; each pump must be able to supply the maximum required capacity for a fire water ring main system. Fire water pumps must be of the submerged vertical type when taking suction from open water and of the horizontal type when suction is taken from a storage tank.

In principle, the pumping system shall

(1) Satisfy the water demand criteria.
(2) Have minimal adverse environmental and community impact.
(3) Comply with environmental requirements.
(4) Comply with OH&S requirements.
(5) Minimize energy consumption by efficient operation.
(6) Have reliable and long service life with minimal maintenance and least whole of life cost.
(7) Provide adequate weather protection and stormwater management.
(8) Provide vehicular and personnel access for maintenance.

The fire water pumps must be installed in a location which is considered to be safe from the effects of fire and clouds of combustible vapor and from collision damage by vehicles and shipping. They should, for example, be at least 100 m away from jetty loading points and from moored tankers or barges handling liquid hydrocarbons. They must be accessible to facilitate maintenance and be provided with hoisting facilities. Figure 5.1 shows a typical fire pump installation.

The main fire water pump must be driven by an electric motor and the second pump, of 100 % standby capacity, by some other power source, preferably a diesel engine. Alternatively, three pumps, each capable of supplying 60 % of the required capacity may be installed, with one pump driven by an electric motor and the other two by diesel engines.

Fire water pumps lie at the heart of a fire-fighting system. Reliable operation under extreme time conditions is a key requirement; therefore, when the required pump capacity exceeds 1,000 m^3, two or more smaller pumps must be installed, together with an adequate number of spare pumps. The power of the drives, for both main and standby units, must be rated, so that it will be possible to start the pumps against an open discharge with pressure in the fire water ring main system under non-fire conditions, normally at 2–3 bar gage unless otherwise agreed by the relevant authorities. The main fire water pump must be provided with automatic starting facilities which will function immediately when the fire alarm system becomes operational due to one of the following actions:

• When a fire call point is operated;
• When an automatic fire detection system is operated;

Fig. 5.1 A fire pump installation

- When the pressure in the fire water ring main system drops below the minimum required static pressure, which is normally 2–3 bar (ga).

  The standby fire water pumps must be provided with automatic starting facilities which will function

- If the main fire water pumps do not start, or having started, fail to build up the required pressure in the fire water ring main system within 20 s.
- Manual starting of each pump unit (without the fire alarms coming into operation) must be possible at the pump, from the control center and, when necessary, from the gate house. Manual stopping of each pump unit must only be possible at the pump.

  For diesel engines, the following additional requirements must also apply:

- The capacity of the fuel tank must be such that the engine can operate on full power for at least 24 h.
- The tank must be installed at a safe distance from the engine, with the bottom at least 0.2 m above the suction valve of the fuel injection pump.

- The tank must be provided with a sump, an expansion dome, a level gage and a low-level alarm which must sound when the level of the fuel has reached the "2 h fuel remaining" level.
- The tank must be provided with facilities and hose connections for refilling directly from drums.

The pumps should have stable characteristic curves exhibiting a decrease in head with increasing capacity from zero flow to maximum flow; a relatively flat curve is preferred with a shutoff pressure not exceeding the design pressure by more than 15 %.

Where technical constraints allow a choice in the type of pumping station arrangement or type of pumping machinery, the final choice will normally be determined as the most cost-effective method. Cost-effectiveness should be determined by a net present value analysis.

Factors to be considered are as follows:

- Cost of pumping station structure and its life.
- Energy cost over the life of the pumping station.
- Maintenance cost and confined space requirements.
- Life and replacement cost of pumping machinery, including ancillary items such as switchgear, lifting gear and ventilation equipment.
- Risk costs.
- Land acquisition cost.
- Net present values of alternatives.

Pumps must take suction from static storage such as tanks, cooling tower basin. The reserve for fire fighting is to be the equivalent of 10 h use at the design figure. Credit may be taken for any incoming makeup water. This reserve for fire fighting should normally be additional to that required for any other user taking water from the same static storage, and the piping arrangements at the storage, where possible, should be so arranged that other users cannot draw on this capacity. The integrity of the makeup water supply must be assured.

The net present value analysis shall allow for the different efficiencies for each suitable pump, the variation in pump duty required for different pipe materials and class of pipe and the economic life of different pipe materials [143–157].

5.5 Winterizing

When the lowest recorded ambient temperature is below 0 °C, water mains must be buried below the frost line, but not less than 0.6 m below ground level. Branches to equipment which must be shutdown while the remainder of the unit is in operation must have the block valves protected by one of the following methods:

- Provide a bypass, just under the block valves, from the supply back to the return. This bypass must be 12.7 mm for lines 76.2 mm and smaller, 25.4 mm for lines

101.6–203 mm and 38 mm for lines larger than 203 mm. Provide a drain in the valve body just above the gate of the valve so that all water can be drained out of the line above the valve after it is closed and another in the bonnet to drain the void around the gate and stem.

- Located in a heated valve box.
- Steam trace the valve.

All portions of water lines above the frost line must be provided with drains. In freezing climate where water lines must be above ground, branch lines must be taken from the top of horizontal main lines with the block valve in a horizontal position and drains must be provided on the dead leg side.

In locations where freezing can occur, the fire water pumps must be installed in housing for protection; for other locations, a rain/sun cover only may be required. When the pump suction is taken from open water, a strainer system which is easy to clean must be provided. When the pump suction is taken from storage, a strainer must be included in the replenishment supply to the storage tank. The discharge line from each pump must be fitted with a check valve, a test valve, a pressure gage and a block valve with a locking device; the test valves must have a common return line with a flow metering unit. Each pump must be connected separately to a common manifold.

The pump common discharge manifold must normally be connected to the fire water ring main system by two separate pipelines each with a block valve and of the same size as the ring main.

In climates where freezing occurs, provisions must be made to prevent stored water from freezing, for example by circulation or by heating; alternatively, storage capacity must be increased to compensate for the ice layer. The quality of the water must be monitored and treated to control the growth of algae and/or barnacles. The replenishment system must also include easy-to-clean strainer facilities [148–159].

5.6 Water Tanks for Fire Protection

Fire water storage tanks have been an important feature of industrial fire protection systems over the last 100 years. The capacity of the tank is the number of cubic meters available above the outlet opening. The net capacity between the outlet opening of the discharge pipe and the inlet of the overflow must be at least equal to the rated capacity. For gravity tanks with large plate risers, the net capacity must be the number of cubic meters between the inlet of the overflow and the designated low-water-level line. For suction tanks, the net capacity must be the number of cubic meters between the inlet of the overflow and the level of the vortex plate.

The location chosen must be such that the tank and structure will not be subject to fire exposure from adjacent units. If lack of yard room makes this impracticable, the exposed steel work must be suitably fireproofed or protected by open sprinklers.

When steel or iron is used for supports inside the building near combustible construction or occupancy, it must be fireproofed inside the building, 1.8 m above combustible roof coverings and within 6.1 m of windows and doors from which fire might issue. Steel beams or braces joining two building columns which support a tank structure must also be suitably fireproofed when near combustible construction or occupancy. Interior timber must not be used to support or brace tank structures.

Fireproofing where necessary must include steel work within 6.1 m of combustible buildings, windows, doors and flammable liquid and gas from which fire might issue.

Fireproofing, where required, must have a fire resistance rating of not less than 2 h.

Foundations or footings must furnish adequate support and anchorage for the tower. If the tank or supporting trestle is to be placed on a building, the building must be designed and built to carry the maximum loads.

Water tanks provide stored water for fire pumps and fire protection systems. Fire water taken from open water is preferred, but if water of acceptable quality for fire fighting in the required quantity cannot be supplied from open water, or if it is not economically justified because of distance to install fire water pumps at an open source, water storage facilities must be provided.

Storage facilities may consist of an open tank of steel or concrete or a basin of sufficient capacity. The tank or basin should have two compartments to facilitate maintenance, each containing 60 % of the total required capacity and there should be adequate replenishment facilities. A single compartment of 100 % capacity is acceptable, providing that an alternative source of water, for example, from temporary storage will be available during maintenance periods. The replenishment rate must normally not be less than 60 % of the total required fire water pumping capacity.

Regardless of domestic use, all tanks need to be equipped with a device that will ensure that the tank contains the designated amount of water for fire flow duration as determined by the fire department. Tank size may be increased to serve multiple structures on a single parcel.

If a 100 % replenishment rate is available, the stored fire water capacity may be reduced if agreed by the Authorities that may be considered for replenishment are plant cooling water, open water or below-ground water, provided that it is available at an acceptable distance and in sufficient quantity for a minimum of 6 h uninterrupted fire fighting at the maximum required rate.

Inspection, testing and maintenance of fire water storage tanks are critical to fire safety. Unless otherwise approved by the relevant authority, the tank must be kept two-thirds full of water, and an air pressure of at least 5.2 bars by the gage must be maintained. As the last of the water leaves the pressure tank, the residual pressure shown on the gage must not be less than zero and must be sufficient to give not less than 1.0 bars pressure at the highest sprinkler under the main roof of the building [149–161]. Figure 5.2 shows a typical water storage tank.

Fig. 5.2 A typical fire fighting water storage tank

5.7 Water Spray–Fixed Systems for Fire Protection

Water remains the predominant fire suppression medium worldwide. By adapting its method of application, water can even be used to extinguish or control fires in situations where it would not normally be considered suitable.

A water spray system is connected to the supply through an automatically or manually actuated flow-control valve. Water is then piped to specially designed nozzles, which distribute it over the protected area.

Water spray is applicable for protection of specific hazards and equipment and may be installed independently of or supplementary to other forms of fire protection systems or equipment.

5.7.1 Hazards

Water spray protection is acceptable for the protection of hazards involving

- Gaseous and liquid flammable materials;
- Electrical hazards such as transformers, oil switches, motors, cable trays and cable runs;
- Ordinary combustibles such as paper, wood and textiles;
- Certain hazardous solids.

In general, water spray may be used effectively for any one or a combination of the following purposes:

- Extinguishment of fire;
- Control of burning;
- Exposure protection;
- Prevention of fire.

The two types of water spray system are as follows:

5.7.2 High-Velocity System

High-velocity (HV) systems are mainly used to extinguish fires involving oils or similar combustible liquids. Oil-filled electrical equipment such as transformers or lubrication systems on steam-driven turbines is generally protected by HV systems.

Extinguishment of fires by HV is achieved by a combination of the following:

- Cooling of the burning oil surface reduces the vaporization rate.
- Steam generated in the fire zone causes a smothering effect through oxygen displacement.
- Dilution of the burning product—a water-miscible product, for example, alcohol can be diluted to a level where it will no longer burn.
- Emulsification through the bombardment of the surface of the product by the high-velocity water droplets. The emulsion formed consists of either globules of oil suspended in water or globules of water suspended in oil. The effect is temporary, and the mixture will separate again in time.

5.7.3 Medium-Velocity Systems

Medium-velocity (MV) systems are mainly used to protect structures, plant and storage vessels from radiated heat and direct flame impingement.

Typical applications include the complete external surface protection of bulk liquefied gas pressure vessels to prevent container failure and the resultant boiling liquid expanding vapor explosion; bulk flammable liquid storage tanks; structures supporting hazardous plant; and equipment such as conveyors.

In addition, fire prevention may be achieved particularly where flammable gas leaks are likely to occur. The action of the spray on the leak will help to dissipate the gas more quickly and prevent concentrations from reaching their flammable limits.

There are limitations to the use of water spray which must be recognized. Such limitations involve the nature of the equipment to be protected, the physical and chemical properties of the materials involved and the environment of the hazard.

Other standards also consider limitations to the application of water (slop over, frothing, electrical clearances, etc.) [143–161].

5.8 Water Supplies

It is of vital importance that water supplies be selected which provide water as free as possible from foreign materials.

The water supply flow rate and pressure must be capable of maintaining water discharge at the design rate and duration for all systems designed to operate simultaneously.

For water supply distribution systems, an allowance for the flow rate of hose streams or other fire protection water requirements must be made in determining the maximum demand.

Sectional control shutoff valves must be located with particular care so that they will be accessible during an emergency.

When only a limited water source is available, sufficient water for a second operation must be provided so that the protection can be reestablished without waiting for the supply to be replenished.

The water supply for water spray systems must be from reliable fire protection water supplies, such as:

• Connections to waterworks systems;
• Gravity tanks (in special cases pressure tanks); or
• Fire pumps with adequate water supply.

5.8.1 Extinguishment

Extinguishment of fires by water spray may be accomplished by surface cooling, by smothering from steam produced, by emulsification, by dilution or by various combinations thereof. Systems must be designed so that within a reasonable period of time, extinguishment must be accomplished and all surfaces must be cooled sufficiently to prevent "flashback" occurring after the system is shutoff.

The design density for extinguishment must be based upon test data or knowledge concerning conditions similar to those that will apply in the actual installation.

A general range of water spray application rates that will apply to most ordinary combustible solids or flammable liquids are from 8.1 $(L/min.)/m^2$ to 20.4 $(L/min.)/m^2$ of protected surface.

Each of the following methods or a combination of them must be considered when designing a water spray system for extinguishment purposes:

• Surface cooling;
• Smothering by steam produced;
• Emulsification;
• Dilution;
• Other factors.

5.9 Fire and Explosion Prevention

The system must be able to function effectively for a sufficient time to dissolve, dilute, disperse or cool flammable or hazardous materials. The possible duration of release of the materials must be considered in the selection of duration times.

The rate of application must be based upon experience with the product or upon test.

Separate fire areas must be protected by separate systems. Single systems must be kept as small as practicable, giving consideration to the water supplies and other factors affecting reliability of the protection. The hydraulically designed discharge rate for a single system or multiple systems designed to operate simultaneously must not exceed the available water supply.

Separation of fire areas must be by space, fire barriers, special drainage or combination of these. In the separation of fire areas, consideration must be given to the possible flow of burning liquids before or during operation of the water spray systems [159–162].

5.9.1 Area Drainage

1. Adequate provisions must be made to promptly and effectively dispose of all liquids from the fire area during operation of all systems in the fire area. Such provisions must be adequate for:

   (a) Water discharged from fixed fire protection systems at maximum flow conditions.
   (b) Water likely to be discharged by hose streams.
   (c) Surface water.
   (d) Cooling water normally discharged to the system.

2. There are three methods of disposal or containment:

   (a) Grading.
   (b) Trenching.
   (c) Underground or enclosed drains.

3. The method used must be determined by:

   (a) The extent of the hazard.
   (b) The clear space available.
   (c) The protection required.

Where the hazard is low, the clear space is adequate, and the degree of protection required is not great; grading is acceptable. Where these conditions are not present, consideration must be given to trenching or underground or enclosed drains [158–163].

Definitions, Terminology and References

Abstraction

The removal of water from any source, either permanently or temporarily, so that it:

(a) ceases to be part of the resources of that area;
(b) is transferred to another source within the area.

Activated Carbon Treatment

A process intended for the removal of dissolved and colloidal organic substances from water and wastewater by absorption on activated carbon; for example, for the amelioration of taste, odor or color.

Aeration

The introduction of air into a liquid.

Aerobic Bacteria; Facultative Anaerobic Bacteria

Bacteria capable of multiplying in either the presence or absence of oxygen.

Aerobic Condition

Descriptive of a condition in which dissolved oxygen is present.

Agglomeration

The coalescence of flocs or particles of suspended matter to form larger flocs or particles which settle or may be caused to float more readily.

Alkalinity

The acid-neutralizing capacity of a water. It is usually expressed as "M" alkalinity (the methyl/orange endpoint at a $pH \approx 4.3$) and "P" alkalinity (the phenolphthalein endpoint at a $pH \approx 8.3$). Several ions contribute to alkalinity, but it is generally due to bicarbonate $(HCO)^{-1}$, carbonate $(CO_3)^{-2}$ and hydroxyl $(OH)^{-1}$ ions.

A. Bahadori et al., *Essentials of Water Systems Design in the Oil, Gas, and Chemical Processing Industries*, SpringerBriefs in Applied Sciences and Technology, DOI: 10.1007/978-1-4614-6516-4, © The Author(s) 2013

Alpha Factor

In an activated sludge plant, the ratio of the oxygen transfer coefficient in mixed liquor to the oxygen transfer coefficient in clean water.

Ammonia Stripping

A method of removing ammoniacal compounds from water by making it alkaline, and aerating.

Aquifer

Water-bearing formation (bed or stratum) of permeable rock, sand or gravel capable of yielding significant quantities of water.

Backwash

That part of the operating cycle of an ion-exchange process wherein a reverse upward flow of water expands the bed, effecting such physical changes as loosening the bed to counteract compacting, stirring up and washing off light-insoluble contaminants to clean the bed, or separating a mixed bed into its components to prepare it for regeneration.

Backwashing

The operation of cleaning a filter with water, or with air and water, by reversing the direction of flow.

Ballasting (Deballasting)

The act of taking on (discharging) ballast water.

Bed Expansion

The effect produced during backwashing; the resin particles become separated and rise in the column. The expansion of the bed due to the increase in the space between resin particles may be controlled by regulating backwash flow.

Biochemical Oxygen Demand (BOD)

The mass concentration of dissolved oxygen consumed under specified conditions by the biological oxidation of organic and/or inorganic matter in water.

Biodegradation

Molecular degradation of organic matter resulting from the complex actions of living organisms, ordinarily in an aqueous medium.

Biofilm (of a Sand Filter)

The film, consisting of living organisms, which forms on the surface of a slow sand filter and which is considered to provide an important part of the effective filtering zone.

Biomass

The total mass of living material in a given body of water.

Biota

The living components of an aquatic system including flora and fauna of a region.

Biotic Index

A numerical value used to describe the biota of a water body, serving to indicate its biological quality.

Blowdown

Blowdown is the continuous or intermittent removal of some of the water in the boiler or cooling water system to reduce concentration of dissolved and/or suspended solids.

BOD

The biological oxygen demand (BOD) water test is used to determine how much oxygen is being used by aerobic microorganism in the water to decompose organic matter.

BOD5

The amount of dissolved oxygen consumed in five days by biological process breaking down organic matter.

Boiler Water

A term construed to mean a representative sample of the circulating boiler water, after the generated steam has been separated, and before the incoming feed water or added chemical becomes mixed with it, so that its composition is affected.

Brackish Water

Water having a dissolved matter content in the range of approximately 1,000–30,000 mg/L.

Break-Point Chlorination

The addition of chlorine to water to the point where free available residual chlorine increases in proportion to the incremental dose of chlorine being added. At this point all of the ammonia has been oxidized.

Breakthrough

The first appearance in the solution flowing from an ion-exchange unit of unadsorbed ions similar to those which are depleting the activity of the resin bed. Breakthrough is an indication that regeneration of the resin is necessary.

Brine

Water having more than approximately 30,000 mg/L of dissolved matter.

Catchment Area

The area draining naturally to a water course or to a given point.

Chemical Oxygen Demand

Chemical oxygen demand (COD) is the equivalent amount of oxygen consumed under specified conditions in the chemical oxidation of the organic and oxidizable inorganic matter contained in a wastewater corrected for the influence of chlorides. In American practice, unless otherwise specified, the chemical oxidizing agent is hot acid dichromate.

Chemical Tracer

A chemical substance added to, or naturally present in water, to allow flow to be followed.

Chemical Treatment

A process involving the addition of chemicals to achieve a specific result.

Chlorine Requirement

The amount of chlorine, expressed in mg/kg, required to achieve under specified conditions the objectives of chlorination.

Chlorine Residual

The amount of available chlorine present in water at any specified period, subsequent to the addition of chlorine.

Coagulation

The precipitation from solution or suspension of fine particles which tend to unite in clots or curds.

Connate Water

Interstitial water of the same geological age as the surrounding rock or bed, often of poor quality and unfit for normal use (for example, potable purposes, industrial and agricultural use).

Cross Connection

A connection between pipes which may cause the transfer of polluted water into a potable water supply with consequent hazard to public health. This term is also used to describe a legitimate connection between different distribution systems.

Decantation

The withdrawal of the supernatant liquor after settlement of suspended solids, or after separation from a liquid of higher density.

Detritus

In a biological context, organic particulate matter. In the context of sewage treatment practice, coarse debris denser than water but capable of being transported in moving water.

Dewatering

The process whereby wet sludge, usually conditioned by a coagulant, has its water content reduced by physical means.

Dissolved Oxygen

Dissolved oxygen (DO) is the oxygen dissolved in sewage, water or other liquid, usually expressed in milligrams per litter or percent of saturation. It is the test used in BOD determination.

Drift

Water lost from a water-cooling tower as liquid droplets entrained in the exhaust air, units: kg per hour or percent of circulating water flow.

Drinking Water/Potable Water

Water of a quality suitable for drinking purposes.

Effluent

Water or wastewater discharge from a containing space such as a treatment plant, industrial process or lagoon.

Feedwater

The water supplied to a boiler to make up for losses.

Fire Hydrant (Underground Fire Hydrant)

An assembly contained in a pit or box below ground level and comprising a valve and outlet connection from a water supply main.

Fire Hydrant Pillar

A fire hydrant whose outlet connection is fitted to a vertical component projecting above ground level.

Floc

Any small, tufted or flake-like mass of matter floating in a solution, for example, as produced by precipitation.

Foam Inlet

Fixed equipment consisting of an inlet connection, fixed piping and a discharge assembly, enabling firemen to introduce foam into an enclosed compartment

Fresh Water

Water having less than approximately 1,000 mg/L of dissolved matter.

Ground Water

Water that fills all of the unlocked of material underlying the water table within the upper limit of saturation.

Hardness

A characteristic of water generally accepted to represent the total concentration of calcium and magnesium ions.

Industrial Water

Any water used for, or during, an industrial process.

Ion Exchange

A chemical process involving the reversible interchange of ions between a solution and a particular solid material (ion exchanger), such as an ion-exchange resin consisting of matrix of insoluble material interspersed with fixed ions of opposite charge.

Langelier's Index

A technique of predicting whether water will tend to dissolve or precipitate calcium carbonate. If the water precipitates calcium carbonate, scale formation may result. If the water dissolves calcium carbonate, it has a corrosive tendency. To calculate Langelier's Index, the actual pH value of the water and Langelier's saturation pH value (pHS) are needed. Langelier's saturation pH value is determined by the relationship between the calcium hardness, the total alkalinity, the total solids concentration and the temperature of the water.

Oxygen Consumed

Oxygen consumed is the quantity of oxygen taken up from potassium permanganate in solution by a liquid containing organic matter commonly regarded as an index of the carbonaceous matter present. Time and temperature must be specified.

Physicochemical Treatment

A combination of physical and chemical treatment to achieve a specific result.

Pollution

Contamination of water with actively or potentially toxic or otherwise harmful materials.

Rain Water

Water arising from atmospheric perception, which has not yet collected soluble matter from the earth.

Resin

A polymer of unsaturated hydrocarbons from petroleum processing, for example, in the cracking of petroleum oils, propane deasphalting, clay treatment of thermally cracked naphthas. Chief uses include:

* rubber and plastics;
* impregnants;
* surface coatings.

Raw Water

Untreated water. Water taken from natural sources, that is, water wells or surface water.

Recirculation Rate

The flow of cooling water being pumped through the entire plant cooling loop.

Regenerant

The solution used to restore the activity of an ion exchanger. Acids are employed to restore a cation exchanger to its hydrogen form; brine solutions may be used to convert the cation exchanger to the sodium form. The anion exchanger may be regenerated by treatment with an alkaline solution.

Rinse

The operation which follows regeneration; a flushing out of excess regenerant solution.

Ryznar Stability Index

An empirical method for predicting scaling tendencies of water based on a study of operating results with water of various saturation indices.

Run-On

Any rain water, leachate or other liquid that drain on to any waste treatment area.

Run-Off

Any rain water, leachate or other liquid that drains over land from any part of a waste treatment facility.

Sampling Point

A specific location on a sampling line at which an individual sample is extracted.

Sedimentation

The process of settling and deposition, under the influence of gravity, of suspended matter carried by water or wastewater.

Self-Purification

The natural processes of purification in a polluted body of water.

Septic Tank

Closed sedimentation tank in which settled sludge is in immediate contact with the wastewater flowing through the tank, and the organic solids are decomposed by anaerobic bacterial action.

Storm Sewage

A mixture of sewage and the surface water arising from heavy rainfall or melting snow (ice).

Storm Water; Storm Water Run-Off

Surface water draining to a watercourse as a result of heavy rainfall.

Sewage Effluent

Treated sewage discharged from a sewage treatment works.

Site

Works or plant where sampling is to be carried out.

Sludge

The accumulated settled solids separated from various types of water as a result of natural or artificial processes.

Softening

The removal of most of the calcium and magnesium ions from water.

Standard Water

Water with known amount of component.

Sterilization

A process which inactivates or removes all living organisms (including vegetative and spore forms) as well as viruses.

Storm Sewage

A mixture of sewage and the surface water arising from heavy rainfall or melting snow (ice).

Stratification

The existence or formation of distinct layers in a body of water identified by thermal or salinity characteristics or by differences in oxygen or nutrient content.

Supply Water

Water, which usually has been treated, that passes into a distribution network or a service reservoir.

Surface Water

Water which flows over, or rests on, the surface of a land mass.

Synthetic Resin

Amorphous, organic, semisolid or solid material derived from certain petroleum oils among other sources; approximating natural resin in many qualities and used for similar purposes.

Threshold Limit Values (TLVs)

This refers to maximum concentration of substances which could be discharged in wastewater issued by the National Standard of Environmental Protection Agency for industrial waste.

Thermocline

The layer in a thermally stratified body of water in which the temperature gradient is at a maximum.

Thickening

The process of increasing the concentration of solids in sludge by the removal of water.

Total Dissolved Solids

Total dissolved solids (TDS) is a measure of the combined content of all inorganic and organic substances contained in a liquid in: molecular, ionized or micro-granular (colloidal sol) suspended form.

Total Organic Carbon (TOC)

TOC is a measure of the amount of carbon in a sample originating from organic matter only. The test is run by burning the sample and measuring the CO_2 produced.

Treated Sewage

Sewage that has received partial or complete treatment for the removal and mineralization of organic and other material.

VOD (Volatile Oxygen Demand)

Compounds which under favorable conditions may participate in photochemical reaction to form oxidants typically exclude methane and ethane.

Wastewater

Water discharged after being used in, or produced by, a process, and which is of no further immediate value to that process.

Wet Chemical Method

Method based on chemical reagent not instrumental.

References

1. T.V. Arden, R.D. Forrest, Water treatment for industrial use. Chem. Eng. **339**, 919–922 (1978)
2. A. Brandelli, M.L. Baldasso, E.P. Goettems, Toxicity identification and reduction evaluation in petrochemical effluents—SITEL case. Water Sci. Technol. **25**(3), 73–84 (1992)
3. A.J.P. Vadovic, Reference file: industrial water treatment systems. Plant Eng. (Barrington, Illinois) **30**(25), 135–138 (1976)
4. A. Bahadori, *Key Technical Points for Process Design of Water Systems* (VDM Verlag Publishing, Saarbrücken, Germany, 2011)
5. American Boiler Manufactures' Association (ABMA)
6. American Society of Mechanical Engineers (ASME)
7. AWWA (American Water Works Association, INC) *Water Treatment Plant Design*, 1971 Manual, M21, "Ground Water"
8. API (American Petroleum Institute) *API Glossary of Terms Used in Petroleum Refining*, 2nd. edn. (1962)
9. GPSA (Gas Processors Suppliers Association) *Engineering Data Book*, 12th edn. (OK, USA 2004)
10. USPHS (US PUBLIC HEALTH SERVICE) *Drinking Water Standards*, No. 956, 1962
11. J.-L. Fernández-Turiel, D. Gimeno, J.-J. Rodriguez, M. Carnicero, F. Valero, Factors influencing the quality of a surface water supply system: the ter river, northeastern Spain. Fresenius Environ. Bull. **12**(1), 67–75 (2003)
12. W. McFarland, Groundwater treatment alternatives for industry. I. Iron and manganese removal. Plant Eng. (Barrington, Illinois) **39**(13), 62–66 (1985)
13. M. Drikas, J.Y. Morran, C. Pelekani, C. Hepplewhite, D.B. Bursill, Removal of natural organic matter—a fresh approach. Water Sci. Technol. Water Supply **2**(1), 71–79 (2002)
14. L.H. Huizar, D. Kang, K. Lansey, World environmental and water resources congress 2011: bearing knowledge for sustainability—Proceedings of the World Environmental and Water Resources Congress (2011), pp. 3238–3250
15. K.K. Kuok, S. Harun, P.-C. Chiu, A review of integrated river basin management for Sarawak River. Am. J. Environ. Sci. **7**(3), 276–285 (2011)
16. A. Unal, N. Tufekci, M. Cakmakci, C. Kinaci, Effect of organic and inorganic matters on the oxidation of Fe(II) in raw water from Ömerli Dam. Desalination Water Treat **26**(1–3), 194–200 (2011)
17. A.S. Krisher, Raw water treatment in the CPI. Chem. Eng. (N.Y.) **85**(19), 79–98 (1978)
18. J. Luo, H.-A. Guan, G.-S. Sun, X.-D. Zhou, Studies on the aeration way of the biological pretreatment process for slightly polluted raw water proceedings of the 2008 global symposium on recycling. Waste Treat Clean Technol REWAS **2008**, 1373–1379 (2008)

A. Bahadori et al., *Essentials of Water Systems Design in the Oil, Gas, and Chemical Processing Industries*, SpringerBriefs in Applied Sciences and Technology, DOI: 10.1007/978-1-4614-6516-4, © The Author(s) 2013

19. G. Sherry, S. Eckols, W. Stauber, Design of underground elements for the raw water intake facilities, city of Austin water treatment plant no. 4. Proceedings—Rapid Excavation and Tunneling Conference (2011), pp. 432–440

20. A.K. Gupta, R.K. Shrivastava, Reliability-constrained optimization of water treatment plant design using genetic algorithm. J. Environ. Eng. **136**(3), 326–334 (2010)

21. A. Bahadori, H.B. Vuthaluru, Prediction of silica carry-over and solubility in steam of boilers using simple correlation. Appl. Therm. Eng. **30**(2010), 250–253 (2010)

22. A. Bahadori, H.B. Vuthaluru, Simple Arrhenius-type function accurately predicts dissolved oxygen saturation concentrations in aquatic systems. Proc. Saf. Environ. Prot. **88**(5), 335–340 (2010)

23. A. Bahadori, H.B. Vuthaluru, S. Mokhatab, Simple correlation accurately predicts aqueous solubility of light alkanes. J. Energy Sources Part A: Recovery, Utilization, Environ. Eff. **31**(9), 761–766 (2009)

24. A. Bahadori, H.B. Vuthaluru, M.O. Tade, S. Mokhatab, Predicting water-hydrocarbon systems mutual solubility. Chem. Eng. Technol. **31**, 1743–1747 (2008)

25. T.R. Camp, *Water and its impurities* (Reinhold Book Corp, New York, 1963), p. 355

26. S.C. Chapra, G.J. Pelletier, *QUAL2 K: a modeling frame work for simulating river and stream water quality: documentation and users manual. Civil and environmental department* (Tufts University, Medford, MA, 2003)

27. F. Civan, Use exponential functions to correlate temperature dependence. Chem. Eng. Prog. **104**, 46–52 (2008)

28. F. Civan, Critical modification to the Vogel—Tammann—Fulcher equation for temperature effect on the density of water ind. Eng. Chem. Res. **46**, 5810–5814 (2007)

29. S.C. Chapra, *Surface water-quality modeling* (McGraw-Hill International Editions, USA, 1997)

30. K.M. Chomicki, S.L. Schiff, Stable oxygen isotopic fractionation during photolytic O2 consumption in stream waters. Sci. Total Environ. **404**, 236–244 (2008)

31. J. Colt, *Computation of dissolved gas concentrations in water as functions of temperature, salinity and pressure, American Fisheries Society publication, 14* (Bethesda, MD, USA, 1984)

32. B.A. Cox (2003) A review of currently available in-stream water-quality models and their applicability for simulating dissolved oxygen in lowland rivers. Sci. Total Environ. pp. 314–316, 335–377

33. W.E. Dobbins, BOD and oxygen relationships in streams. J. Sanit Eng. Div. ASCE **90**(SA3), 53–78 (1964)

34. R.L. Droste, *Theory and practice of water and wastewater treatment* (Wiley, NY, USA, 1997)

35. R.W. Edwards, M. Owens, The oxygen balance of streams. Ecology and the industrial society, Fifth Symposium of the British Ecological Society (Blackwell, Oxford, 1965), pp. 149–172

36. G.S. Fulcher, Analysis of recent data of the viscosity of glasses. J. Am. Ceram. Soc. **8**, 339–355 (1925)

37. A. James, An introduction to water quality modelling, 2nd edn. (Wiley, USA, 1993)

38. P.R. Kannel, S. Lee, Y.S. Lee, S.R. Kanel, G.J. Pelletier, Application of automated QUAL2Kw for water quality modeling and management in the Bagmati river. Nepal. Ecol. Model. **202**, 503–517 (2007)

39. D. Kim, Q. Wang, G.A. Soriala, D.D. Dionysioua, D. Timberlakeb, A model approach for evaluating effects of remedial actions on mercury speciation and transport in a lake system. Sci. Total Environ. **327**, 1–15 (2004)

40. C. Neal, N. Christophersen, R. Neale, C.J. Smith, P.G. Whitehead, B. Reynolds, Chloride in precipitation and streamwater for the upland catchment of river Severn, mid-wales: some consequences for hydrochemical models. Hydrol. Process. **2**, 155–165 (1988)

41. M. Owens, R. Edwards, Gibbs J. Some reaeration studies in streams. Int. J. Air Water Pollut. **8**, 469–486 (1964)
42. V. Palmieri, R.J. de Carvalho, Qual2e model for the Corumbatai River. Ecol. Model. **198**, 269–275 (2006)
43. S.S. Park, Y.S. Lee, Water quality modeling study of the Nakdong river, Korea. Ecol. Model. **152**, 66–75 (2002)
44. N.E. Peters, E.B. Ratcliffe, M. Tranter, Tracing solute mobility at the Panola mountain research Watershed, Georgia: variations in Naq, Cly and H4SiO4 concentrations. Hydrol Water Res and Ecol. In: Headwaters, Proceedings of the Headwater '98 Conference, vol. 248 (IAHS Publication, Meranr/Merano, Italy, 1998), 483–490
45. J.O. Reuss, D.W. Johnson, Acid deposition and the acidification of soils and waters. Ecol. Stud. **59**, 119 (1986)
46. R. Smart, C.C. White, J. Townend, M.S. Cresser, A model for predicting chloride concentrations in river water in a relatively unpolluted catchment in north-east Scotland. Sci. Total Environ. **265**, 131–141 (2001)
47. C. Soulsby, Influence of sea-salt on stream water chemistry in an upland afforested catchment. Hydrol. Process. **9**, 183–196 (1995)
48. Standard method for the examination of water and wastewater, 18 edn., APHA, AWWA and WEF, (American Public Health Association, Washington DC, USA, 1992)
49. M. Streat, K. Hellgardt, N.L.R. Newton, Hydrous ferric oxide as an adsorbent in water treatment: Part 2. Adsorption studies. Proc. Saf. Environ. Prot. **86**(1), 11–20 (2008)
50. R.G. Wetzel, *Limnology: lake and river ecosystems*, 3rd edn. (Academic Press, San Diego, USA, 2001)
51. R. Rautenbach, K. Voßenkaul, Pressure driven membrane processes—the answer to the need of a growing world population for quality water supply and waste water disposal. Sep. Purif. Technol. **22–23**, 193–208 (2001)
52. D.W. Hendricks, *Fundamentals of water treatment unit processes: physical, chemical, and biological* (IWA Publishing, UK, 2010)
53. J.-Q. Jiang, Development of coagulation theory and new coagulants for water treatment: Its past, current and future trend. Water Sci. Technol. Water Supply **1**(4), 57–64 (2001)
54. T. Leiknes, The effect of coupling coagulation and flocculation with membrane filtration in water treatment: a review. J. Environ. Sci. **21**(1), 8–12 (2009)
55. S. Haydar, H. Ahmad, J.A. Aziz, Optimization of coagulation-flocculation in the treatment of canal water. Environ. Eng. Manage. J. **9**(11), 1563–1570 (2010)
56. K. Konieczny, M. Rajca, M. Bodzek, A. Kwiecińska, Water treatment using hybrid method of coagulation and low-pressure membrane filtration. Environ. Prot. Eng. **35**(1), 5–22 (2009)
57. T.K. Trinh, L.S. Kang, Response surface methodological approach to optimize the coagulation-flocculation process in drinking water treatment. Chem. Eng. Res. Des. **89**(7), 1126–1135 (2011)
58. D.L. Gone, J.-L. Seidel, C. Batiot, K. Bamory, R. Ligban, J. Biemi, Using fluorescence spectroscopy EEM to evaluate the efficiency of organic matter removal during coagulation-flocculation of a tropical surface water (Agbo reservoir). J. Hazard. Mater. **172**(2–3), 693–699 (2009)
59. M.E. Walsh, N. Zhao, S.L. Gora, G.A. Gagnon, Effect of coagulation and flocculation conditions on water quality in an immersed ultrafiltration process. Environ. Technol. **30**(9), 927–938 (2009)
60. S. Kawamura, Optimisation of basic water-treatment processes—design and operation: coagulation and flocculation. J. Water Supply Res. Technol.AQUA **45**(1), 35–47 (1996)
61. H. Bernhardt, H. Schell, B. Lusse, Criteria for the control of flocculation and filtration processes in the water treatment of reservoir water. Water Supply **4**(4), 99–116 (1986)
62. C.R. O'Melia, Coagulation and sedimentation in lakes, reservoirs and water treatment plants. Water Sci. Technol. **37**(2), 129 (1998)

63. W.C.G. Ko, B.W.L. Mak, S.S.K. Pun, Pilot testing of manganese oxidation in a high-rate sedimentation process for the reprovisioning of the Sha Tin water treatment works. Hong Kong Spec. Adm. Reg. Water Environ. J. **21**(1), 26–33 (2007)

64. A.I. Zhukova, E.A. Kvasnitsa, G.N. Malysh, Z.G. Ivanova, V.T. Ostapenko, Y.I. Tarasevich, Reagent treatments of sediments built up in sedimentation tanks of the Dnieper river water station. J. Water Chem. Technol. **32**(6), 348–351 (2010)

65. G. Joh, Y.S. Choi, J.-K. Shin, J. Lee, Problematic algae in the sedimentation and filtration process of water treatment plants. J. Water Supply Res. Technol. AQUA **60**(4), 219–230 (2011)

66. W.J. Voortman, I.W. Bailey, C.D. Reddy, G.E. Rencken, Disinfection methods for package water treatment plants. Water Supply **13**(2), 203–218 (1995)

67. J.-C. Lou, Y.-C. Lin, Treatment efficiency and formation of disinfection by products in advanced water treatment process. Environ. Eng. Sci. **25**(1), 82–91 (2008)

68. H.Z. Zastawny, H. Romat, N.K. Leitner, J.S. Chang, Pulsed arc discharges for water treatment and disinfection. Inst. Phys. Conf. Ser. **178**, 325–330 (2004)

69. M.R. Teixeira, S.M. Rosa, V. Sousa, Natural organic matter and disinfection by-products formation potential in water treatment. Water Res. Manag. **25**(12), 3005–3015 (2011)

70. D. Bixio, I. Boonen, C. Thoeye, G. De Gueldre, Experience with phosphorus removal and sludge handling and disposal in Flanders. Water Sci. Technol. **52**(4), 19–25 (2005)

71. R. Saunamaki, Sludge handling and disposal at Finnish activated sludge plants. Water Sci. Technol. **20**(1), 171–182 (1988)

72. M.P.J. Weemaes, W.H. Verstraete, Evaluation of current wet sludge disintegration techniques. J. Chem. Technol. Biotechnol. **73**(2), 83–92 (1998)

73. D.R. Argo, Use of lime clarification and reverse osmosis in water reclamation. J. Water Pollut. Control Fed. **56**(12), 1238–1246 (1984)

74. K.E. Dennett, R.T. Dixon, Using carbon dioxide to cope with fluctuations in raw water pH and maintain effective conventional treatment. J. Water Supply Res. Technol. AQUA **52**(5), 369–381 (2003)

75. X. Zhang, S. Zhang, Y. Liu, The treatment of algae-laden raw water with compact flofilter of dissolved air flotation and GAC deep bed filtration. Water Sci. Technol. Water Supply **4**(5–6), 35–41 (2004)

76. Mark R. Wiesner, Pierre Mazounie, Raw water characteristics and the selection of treatment configurations for particle removal. J. Am. Water Works Assoc. **81**(5), 80–89 (1989)

77. J.E. Dyksen, S. Master, V. Veerapaneni, Treatment of the Hudson River for water supply development. J. New England Water Works Assoc. **122**(3), 253–258 (2008)

78. T. Oe, H. Koide, H. Hirokawa, K. Okukawa, Performance of membrane filtration system used for water treatment. Desalination **106**(1–3), 107–113 (1996)

79. A. Huq, B. Xu, M.A.R. Chowdhury, M.S. Islam, R. Montilla, R.R. Colwell, A simple filtration method to remove plankton-associated Vibrio cholerae in raw water supplies in developing countries. Appl. Environ. Microbiol. **62**(7), 2508–2512 (1996)

80. S. Kawamura, Optimisation of basic water-treatment processes-design and operation: sedimentation and filtration. J. Water Supply Res. Technol. AQUA **45**(3), 130–142 (1996)

81. G. Heinicke, F. Persson, W. Uhl, M. Hermansson, T. Hedberg, The effect of biological pre-filtration on the performance of conventional surface water treatment. J. Water Supply Res. Technol. AQUA **55**(2), 109–119 (2006)

82. X.-Q. Chu, J. Kim, K. Park, S. Choi, H.-S. Kim, Coagulation pre-treatment to reduce membrane fouling in the microfiltration process of Nakdong River water. Water Sci. Technol. Water Supply **9**(4), 357–367 (2009)

83. Y. Chen, G.-Z. Zhang, Z.-W. Xiao, Yangtze River water treatment by spiral UF membrane with coagulation pretreatment. Adv. Mater. Res. **113–116**, 1524–1528 (2010)

84. F.H. Frimmel, F. Saravia, A. Gorenflo, NOM removal from different raw waters by membrane filtration. Water Sci. Technol. Water Supply **4**(4), 165–174 (2004)

85. E. Yüksel, Ö. Akgiray, E. Soyer, Direct filtration with preozonation for small water treatment systems. Water Sci. Technol. **48**(11–12), 473–479 (2004)
86. P. Avery, *Sediment control at intakes: a design guide* (BHRA Fluid Engineering, Bedford, UK, 1989)
87. H. Lauterjung, G. Schmidt, *Planning of intake structures* (Vieweg, Braunschweig, Germany, 1989)
88. C. Zani, D. Feretti, A. Buschini, P. Poli, C. Rossi, Toxicity and genotoxicity of surface water before and after various potabilization steps. Mutat. Res. **587**, 26–37 (2005)
89. Bureau of Reclamation, *Design of small dams* (Bureau of Reclamation, Washington, DC, USA, 1973)
90. J.L. Wood, J. Richardson, *Design of small water storage and erosion control dams* (Colorado State University, Fort Collins, CO, USA, 1975). 263
91. D.L. Boccelli, M.J. Small, U.M. Diwekar, Treatment plant design for particulate removal: effects of flow rate and particle characteristics. J. Am. Water Works Association **96**(11), 77–90 + 12 (2004)
92. J.D. Eisnor, G.A. Gagnon, Assessment of process design and operation in membrane water treatment plants. Annual Conference Abstracts—Canadian Society for Civil Engineering (2000), p. 167
93. C. Zhou, N. Gao, W. Wang, W. Chu, Productive experiment of Huangpu River raw water. J. Huazhong Univ. Sci. Technol. (Natural Science Edition) **39**(7), 128–132 (2011)
94. J. Dahlquist, M. Kulesza, Pre-treatment with dissolved air flotation considering an integrated process design. Water Sci. Technol. Water Supply **1**(2), 115–122 (2001)
95. E. Hadizadeh, K.H. Shirazi, Dynamical modeling for improvement of water treatment system using bond graph method. Proceedings of the IASTED International Conference on Modelling and Simulation (2011), pp. 116–122
96. H. Jan, Cooperative financing for an emergency supply source Peekskill/Montrose raw water interconnect Wines. J. New Engl. Water Works Assoc. **111**(1), 61–65 (1997)
97. C.R. O'Melia, Particles, pretreatment, and performance in water filtration. J. Environ. Eng. **111**(6), 874–890 (1985)
98. S.B. Kwon, H.W. Ahn, C.J. Ahn, C.K. Wang, A case study of dissolved air flotation for seasonal high turbidity water in Korea. Water Sci. Technol. **50**(12), 245–253 (2004)
99. J.G. Janssens, A. Buekens, Assessment of process selection for particle removal in surface water treatment. Aqua **42**(5), 279–288 (1993)
100. B.C. Vallance, R.G. Pritchard, E.E. Hargesheimer, R.T. Seidner, Consideration of DAF retrofit to a large conventional water treatment plant. Water Sci. Technol. **31**(3–4), 93–101 (1995)
101. H.-W. Ahn, N.-S. Park, S. Kim, S.-Y. Park, C.-K. Wang, Modeling of particle removal in the first coarse media of direct horizontal-flow roughing filtration. Environ. Technol. **28**(3), 339–353 (2007)
102. J. Heinanen, P. Jokela, T. Ala-Peijari, Use of dissolved air flotation in potable water treatment in Finland. Water Sci. Technol. **31**(3–4), 225–238 (1995)
103. Hardam Singh Azad, Industrial wastewater management handbook (1976)
104. A. Khatoonabadai, A.R.M. Dehcheshmeh, Oil pollution in the Caspian Sea coastal waters. Int. J. Environ. Pollut. **26**(4), 347–363 (2006)
105. S. Anisuddin, N. Al-Hashar, S. Tahseen, Prevention of oil spill in seawater using locally available materials. Arab. J. Sci. Eng. 30 (2005)
106. R.W. Ladd, D.D. Smith, System study of oil spill cleanup procedure. Paper Number SPE 3047-MS. Presented at Fall Meeting of the Society of Petroleum Engineers of AIME, Houston, Texas, 4–7 October 1970
107. A.A. Al-Majed, A.R. Adebayo, M.E. Hossain, A sustainable approach to controlling oil spills. J. Environ. Manage. **113**, 213–227 (2012)

108. I. Nilssen, S. Johnsen, Holistic environmental management of discharges from the oil and gas industry—combining quantitative risk assessment and environmental monitoring, society of petroleum engineers. 9th International Conference on Health, Safety and Environment in Oil and Gas Exploration and Production 2008—"In Search of Sustainable Excellence" 2, (2008), pp. 606–614

109. A. Fakhru'l-Razi, A. Pendashteh, L.C. Abdullah, D.R.A. Biak, S.S. Madaeni, Z.Z. Abidin, Review of technologies for oil and gas produced water treatment. J. Hazard. Mater. **170**(2–3), 530–551 (2009)

110. J.T. Tanacredi, Petroleum hydrocarbons from effluents: detection in marine environment. J. Water Pollut. Control Fed. **49**(2), 216–226 (1977)

111. C.E. Emole, Regulation of oil and gas pollution. Environ. Policy Law **28**(2), 103–112 (1998)

112. D.W. Dixon-Hardy, S. Beyhan, I.G. Ediz, K. Erarslan, The use of oil refinery wastes as a dust suppression surfactant for use in mining. Environ. Eng. Sci. **25**(8), 1189–1195 (2008)

113. M. Coupard, R. Hournac, New processing trends for reducing oil refinery pollution. Ind. Environ. **8**(2), 26–30 (1985)

114. R.A. Bayoumi, A.Y. El-Nagar, Safe control methods of petroleum crude oil pollution in the mangrove forests of the Egyptian Red sea coast. J. Appl. Sci. Res. **5**(12), 2435–2447 (2009)

115. R.L. Knight, R.H. Kadlec, H.M. Ohlendorf, The use of treatment wetlands for petroleum industry effluents. Environ. Sci. Technol. **33**(7), 973–980 (1999)

116. C. D'Unger, D. Chapman, R.S. Carr, Discharge of oilfield-produced water in Nueces Bay, Texas: A case study. Environ. Manage. **20**(1), 143–150 (1996)

117. J.V. Headley, Y. Gong, S.L. Barbour, R.W. Thring, Evaluation of the ideality of benzene, toluene, ethylbenzene, and xylene on activity coefficients in gas condensate and the implications for dissolution in groundwater. Can. Water Res. J. **25**(1), 67–79 (2000)

118. C.E. Baukal, R. Hayes, M. Grant, P. Singh, D. Foote, Nitrogen oxides emissions reduction technologies in the petrochemical and refining industries. Environ. Prog. **23**(1), 19–28 (2004)

119. W. Parker, G.J. Farquhar, Treatment of a petrochemical wastewater in an anaerobic packed bed reactor. Water pollut. Res. J. Can. **24**(2), 195–205 (1989)

120. J. Li, A GIS planning model for urban oil spill management. Water Sci. Technol. **43**(5), 239–244 (2001)

121. M.F.N. Abowei, 1996 Prediction and consequences of petroleum spills into the Nigerian aquatic environment in the year. Int. J. Environ. Pollut. **6**(2–3), 306–321 (2000)

122. W.-Y. Chiau, Changes in the marine pollution management system in response to the Amorgos oil spill in Taiwan. Mar. Pollut. Bull. **51**(8–12), 1041–1047 (2005)

123. P.D. Lundegard, P.C. Johnson, Source zone natural attenuation at petroleum hydrocarbon spill sites—II: application to a former oil field. Ground Water Monit. Rem. **26**(4), 93–106 (2006)

124. H. Bi, D. Rissik, M. MacOva, L. Hearn, J.F. Mueller, B. Escher, Recovery of a freshwater wetland from chemical contamination after an oil spill. J. Environ. Monit. **13**(3), 713–720 (2011)

125. I. Keramitsoglou, C. Cartalis, P. Kassomenos, Decision support system for managing oil spill events. Environ. Manage. **32**(2), 290–298 (2003)

126. X. Liu, K.W. Wirtz, A. Kannen, D. Kraft, Willingness to pay among households to prevent coastal resources from polluting by oil spills: a pilot survey. Mar. Pollut. Bull. **58**(10), 1514–1521 (2009)

127. K.F. Chen, C.M. Kao, J.Y. Wang, T.Y. Chen, C.C. Chien, Natural attenuation of MTBE at two petroleum-hydrocarbon spill sites. J. Hazard. Mater. **125**(1–3), 10–16 (2005)

128. C. Zhu, L. Wu, S. Li, Application of combined matlab and VB model in water pollution control planning. Key Eng. Mater. **439–440**, 407–410 (2010)

129. B. Crabtree, A.J. Seward, L. Thompson, A case study of regional catchment water quality modelling to identify pollution control requirements. Water Sci. Technol. **53**(10), 47–54 (2006)
130. Y. Liu, W.-H. Zhu, The way to control water pollution of the Huaihe River. Water Sci. Technol. **26**(9–11), 2559–2562 (1992)
131. E.J. Hoffman, Oil spills in Narragansett Bay: comparison between federal and state records. Mar. Pollut. Bull. **16**(6), 240–243 (1985)
132. J.M. Price, M. Reed, M.K. Howard, W.R. Johnson, Z.-G. Ji, C.F. Marshall, N.L. Guinasso Jr, G.B. Rainey, Preliminary assessment of an oil-spill trajectory model using satellite-tracked, oil-spill-simulating drifters. Environ. Model. Softw. **21**(2), 258–270 (2006)
133. X. Liu, K.W. Wirtz, The economy of oil spills: Direct and indirect costs as a function of spill size. J. Hazard. Mater. **171**(1–3), 471–477 (2009)
134. A.H. Al-Rabeh, H.M. Cekirge, N. Gunay, Modeling the fate and transport of Al-Ahmadi oil spill. Water Air Soil Pollut. **65**(3–4), 257–279 (1992)
135. Z. Wang, M. Fingas, S. Blenkinsopp, G. Sergy, M. Landriault, L. Sigouin, P. Lambert, Study of the 25-year-old Nipisi oil spill: persistence of oil residues and comparisons between surface and subsurface sediments. Environ. Sci. Technol. **32**(15), 2222–2232 (1998)
136. S. Chalmers, A. Kowse, P. Stark, L. Facer, N. Smith, Treatment of coal seam gas water. Water **37**(4), 71–76 (2010)
137. F. van Bergen, P. Krzystolik, N. van Wageningen, H. Pagnier, B. Jura, J. Skiba, P. Winthaegen, Z. Kobiela, Production of gas from coal seams in the Upper Silesian Coal Basin in Poland in the post-injection period of an ECBM pilot site. Int. J. Coal Geol. **77**(1–2), 175–187 (2009)
138. C.R. Johnston, G.F. Vance, G.K. Ganjegunte, Irrigation with coalbed natural gas co-produced water. Agric. Water Manag. **95**(11), 1243–1252 (2008)
139. F.Y.C. Huang, P. Natrajan, Feasibility of using natural zeolites to remove sodium from coal bed methane-produced water. J. Environ. Eng. **132**(12), 1644–1650 (2006)
140. American Water Works Association, Distribution System Requirements for Fire Protection, AWWA manual M31, 3rd edn (1998)
141. V. Babrauskas, Fire modelling tools for FSE: are they good enough. J. Fire Prot. Eng. **8**(2), 87–96 (1996)
142. V. Babrauskas, R.D. Peacock, Heat release rate: the single most important variable in fire hazard. Fire Saf. J. **18**, 255–272 (1992)
143. C.B. Edwards, Quantitative fire performance, testing of compressed air foam. Fire Engineering (July 1998), pp. 119–122
144. D.D. Drysdale, An introduction to fire dynamics, (Wiley, New York, USA, 1985)
145. R.M. Clark, H.G. Goddard, Cost and quality of water supply. Am. Water Works Assoc. J. **69**(1), 13–15 (1977)
146. Z. Li, S. Zhou, Study of flow characteristic and fire-fighting performance of viscous water extinguishant. Proc. Saf. Sci. Technol. Part B **3**, 1268–1271 (2002)
147. X.-Q. Liu, H. Gong, Study of fire fighting system to extinguish full surface fire of large scale floating roof tanks Lang. Procedia Eng. **11**, 189–195 (2011)
148. S. Svensson, A study of tactical patterns during fire fighting operations. Fire Saf. J. **37**(7), 673–695 (2002)
149. H. Yu, Design of steady high-pressure fire-fighting water supply system for petrochemical enterprises. Petrol. Refinery Eng. **33**(2), 13–15 (2003)
150. M. Rigolio, Fire fighting in process plants analysis of actual firewater consumption Chemical. Eng. Trans. **19**, 297–302 (2010)
151. C.R. Theobald, The design of a general purpose fire-fighting jet and spray branch. Fire Saf. J. **7**(2), 177–190 (1984)
152. M. Rigolio, Fire fighting in process plants analysis of actual firewater consumption Chemical. Eng. Trans. **19**, 297–302 (2010)

153. J.F. Widmann, J. Duchez, The effect of water sprays on fire fighter thermal imagers. Fire Saf. J. **39**(3), 217–238 (2004)
154. B. Haight, Focusing on the fire station. Diesel Prog. North Am. Edn. **68**(10), 20–25 (2002)
155. A.P. Hatton, M.J. Osborne, The trajectories of large fire fighting jets. Int. J. Heat Fluid Flow **1**(1), 37–41 (1979)
156. Y. Xinmin, H. Feixue, Y. Xinbin, Deploying fire trucks and water sources. Fire Technol. **35**(2), 179–183 (1999)
157. J. Stephens, Improving sprinkler fire control with lower water pressures. Fire Prev. **257**, 32–36 (1993)
158. R.W. Rivard, Cross connection control's role in protecting water quality: fire protection systems. J. New Engl. Water Works Assoc. **111**(3), 304–307 (1997)
159. M.B. Kim, Y.J. Jang, M.O. Yoon, Extinction limit of a pool fire with a water mist. Fire Saf. J. **28**(4), 295–306 (1997)
160. G.V. Hadjisophocleous, J.K. Richardson, Water flow demands for fire fighting. Fire Technol. **41**(3), 173–191 (2005)
161. V. Novozhilov, B. Moghtaderi, D.F. Fletcher, J.H. Kent, Numerical simulation of enclosed gas fire extinguishment by a water spray. J. Appl. Fire Sci. **5**(2), 135–146 (1996)
162. D. Torvi, G. Hadjisophocleous, M.B. Guenther, G. Thomas, Estimating water requirements for fire fighting operations using FIERA system. Fire Technol. **37**(3), 235–262 (2001)
163. J.G. Leibman, K.L. Conover, Fire pumps and controllers. Plumbing Eng. **17**(4), 19–22 (1989)